# The Polar Bear

# THE POLAR BEAR

## Ian Stirling

## Photographs by Dan Guravich

BLANDFORD

First published in the UK 1990 by
Blandford Press, an imprint of Cassell Plc,
Artillery House, Artillery Row, London SW1P 1RT

First published in the United States of America by
The University of Michigan Press

Distributed in Australia by
Capricorn Link (Australia) Pty Ltd,
PO Box 665, Lane Cove, NSW 2066

**British Library Cataloguing in Publication Data**
Stirling, Ian
  The polar bear.
  1. Polar bears
  I. Title  II. Guravich, Dan
  599.74'446

  ISBN 0-7137-2194-4

Typeset in the United States of America
Printed and bound in Japan

*To my parents,*
Margaret and Andrew,
*for their continued understanding and support*
*of an inquiring mind and an independent spirit*

# Acknowledgments

Over the years, I have had a tremendous amount of support from a large number of organizations and people without whom I could not have done the research that enabled me to write this book. The Canadian Wildlife Service has provided steady support and allowed me the freedom to conduct a wide range of studies. The Polar Continental Shelf Project, which is nothing short of the most phenomenal organization in the Canadian Arctic today, constantly supported several of my projects, which could not have been completed without them. I would also like to thank the following organizations for their support: Natural Sciences and Engineering Research Council, Department of Zoology and Boreal Institute of the University of Alberta, Manitoba Department of Natural Resources, Northwest Territories Wildlife Service, Department of Fisheries and Oceans, Department of Indian and Northern Affairs, Dome Petroleum Limited, Esso Resources Canada Limited, and Petro Canada Limited.

I have also been extremely fortunate in the people that I have had to work with. Dennis Andriashek and Wendy Calvert, who have worked with me for many years, deserve special thanks. Their unflagging dedication in the field and in the lab, their constant efforts to do things thoroughly and accurately, and their friendship have been more help than I could describe. I also owe a special debt of thanks to my friend and colleague, Tom Smith, with whom I have studied polar bears and seals all over the Arctic and from whom I have learned a tremendous amount during many long discussions, over too many cups of coffee, during bad weather days. Several other colleagues, graduate students, and assistants have contributed enormously to the field and lab work, and to my appreciation of many things to do with polar bears and their ecological relationships. These include Steve Amstrup, Ralph Archibald, Doug DeMaster, Andy Derocher, V. A. Hughey, Chuck Jonkel, Steve Kearney, Michael Kingsley, Paul Latour, Thor Larsen, John Lee, Nick Lunn, Sam Miller, Nils Øritsland, Malcolm Ramsay, Ray Schweinsburg, Merlin Shoesmith, Don Siniff, Becky Sjare, Pauline Smith, Cheryl Spencer, Gordon Stenhouse, Mitch Taylor, Ian Thorliefson, Savva Uspenski, and Christian Vibe. We have flown with innumerable excellent and hardworking pilots, but Gene Burelson, Ed Long, S. I. "Kobi" Kobiyashi, and Steve Miller (who has now helped catch more polar bears than any other pilot in the world) each did several contracts and were a pleasure to work with. I would also like to thank the following Inuk hunters, from whom I learned a great deal in the field or in discussion: Andy Carpenter, the late Fred Carpenter, Albert Elias, Peter Esau, Ipeelie Inookie,

John Lucas, Wallace Lucas, David Nasogaluak, Henry Nasogaluak, Jimmy Memorana, Fred Wolkie, Geddes Wolkie, and the late Jim Wolkie.

I am very grateful to Downs Matthews for his careful editing of the entire text, and to Susan Popowich for drawing the figures.

Last but not least I would particularly like to thank my wife, Stella, and my children, Lea, Claire, and Ross, for their constant interest and support. I have been away from them a good deal while doing the work described in the book but they have understood.

# Contents

# Figures

# Tables

# Introduction

The polar bear, more than any other animal, symbolizes the Arctic. All of us, whether we are ever likely to see a polar bear or not, know what it looks like. Wherever it goes, Nanuk (its Inuit name) commands respect. Like the surrounding landscape, the polar bear is impressive in its sheer size and rugged beauty. It is both beautiful and potentially dangerous. Its pristine whiteness matches the backdrop of snow and ice that we all associate with the Arctic.

Unlike the caribou or snow geese that take our breath away with their sheer numbers, it is the solitary nature of this arctic nomad that captures our imagination. The polar bear stands alone, an entity, complete in itself.

To me, the wild polar bear is the Arctic incarnate. When watching one amble across the pack ice, looking about and periodically sniffing the wind, there is an overwhelming sense that it belongs there. The Arctic is not a forsaken wasteland to a polar bear; it is home, and a comfortable home at that. For thousands of years, the climate, the ice, and the seals upon which it feeds have shaped and finely tuned the evolution of this predator so exquisitely that it has become not just a symbol but the very embodiment of life in the Arctic.

From the time the polar bear began to separate from its terrestrial brown bear ancestors and venture out onto the sea ice, its links to the land began to diminish. In time, it developed into the truly maritime bear that its latin name, *Ursus maritimus*, would suggest. Polar bears now live on the sea ice for the maximum amount of time, but sometimes it is still necessary to return to land. For example, most females have their maternity dens in snowdrifts on land along the coastline of the continents or polar archipelagoes.

Although the polar bear is a true marine mammal in the sense that it depends on the sea for existence, the fact that it walks about like a land bear, and periodically comes ashore, confuses some. For example, in the United States, the polar bear is considered a marine mammal for legal purposes while in Canada it is a land mammal. Ecologically, however, it is clearly an integral part of the marine ecosystem and that is the context in which I will treat it.

I started my research on polar bears in 1970 after spending several years studying seals in Antarctica, Australia, and elsewhere. International concern for polar bears was high and the Polar Bear Specialists Group of the International Union for the Conservation of Nature (IUCN) had already begun holding biennial meetings that led to the International Agreement on the Conservation of Polar Bears and Their Habitat and to exchange information. A small corps of dedicated scientists maintained continuity in polar bear studies through those early years and set the groundwork for the great expansion in research that followed in the 1970s. They were Jack Lentfer (Alaska), Thor Larsen (Norway), Chuck Jonkel (Canada), Christian Vibe (Denmark), and Savva Uspenski (Union of Soviet Socialist Republics).

In the mid- to late 1960s, polar bear scientists concentrated on determining the sizes of polar bear populations in different areas of the world. The animals were being hunted, and management plans were urgently needed. To acquire this information, scientists from all five "polar bear countries" devised such techniques as drugging bears with tranquilizer darts fired from helicopters. Studies of population size began in Svalbard, Alaska, and Hudson Bay in Canada.

Fairly early on, it became clear the polar bear was not threatened with extinction. Nevertheless, the population was vulnerable to serious population decline if it was not properly managed. In recognition of this, the Polar Bear Specialists Group began to recommend more fundamental research and stressed the importance of broadly based studies on the ecology of the bears.

I was extremely fortunate in that the Canadian Wildlife Service allowed me a lot of latitude in what to do research on. As long as my project was fully meeting its responsibilities in the management-oriented area, we could also conduct behavioral and ecological studies to enhance our overall understanding of the biology of the polar bear. As noted in the Acknowledgments, several other organizations gave long-term support to projects in many related areas and helped make this possible. As an Adjunct Professor in the Zoology Department at the University of Alberta, I involved graduate students in following up specific areas and sharing in the excitement of doing original research in the Arctic. The opinions expressed in the book are my own, however, and do not necessarily reflect those of the Canadian

Wildlife Service, the University of Alberta, or any of the agencies that have supported my research.

Satisfactory management of an animal such as the polar bear requires an understanding of its behavior and activity patterns. For example, how do bears respond to large-scale changes between years in the distribution of sea ice or the seals they feed upon? How many seals do they catch each day? How many cubs do they have each year? How well do cubs survive and why do some die? What is the annual energy requirement of a pregnant bear? What factors determine the distribution of maternity dens? How might offshore drilling for oil affect polar bears? Studies on these topics and others are continuing to shape the conservation and management of polar bears on an ongoing basis.

The genesis of this book came from another area of deep personal interest, that of education and public information. For years, my colleagues and I have assisted photographers, filmmakers, and writers so they could tell the public about polar bears. I have given countless talks to groups as diverse as elementary schools, service clubs, university courses, and professional meetings, and continue to do so. I never cease to be amazed and pleased at the high level of public interest in polar bears, and how many questions people ask. Most of the questions have fairly straightforward answers, but the information is not easily acquired. Much of it is buried in scientific journals and written in technical jargon. Several people suggested that I make the facts more readily available by writing a popular but scientifically accurate book. I liked the idea but always seemed to be too busy completing technical reports. Finally, a friend said to me, "When are you going to write something about polar bears that you don't need a university degree to read?" He was right. It was time.

Shortly after that, I was chatting with Dan Guravich in, appropriately, the Churchill, Manitoba, airport and he made a similar comment.

Knowing he had photographed polar bears more extensively than anyone else in the world, and that he was also interested in educaton, I responded that I was ready to begin. I asked him if he would like to supply the photographs and he quickly agreed.

In writing the book, I have followed a couple of general themes. The first of these is that the polar bear does not exist in isolation. It is both a product and a part of the polar marine ecosystem. It is constantly influenced by a changing environment and it interacts with other species on a daily basis. The polar bear has been a significant factor in the evolution of the behavior and ecology of the arctic seals and vice versa. The arctic fox has learned to follow the bears out on the sea ice during the winter and survive by scavenging on the remains of seal kills. More recently, in an evolutionary sense, the white bear has played a significant role in the culture and spiritual beliefs of the indigenous people that lived around the rim of the polar basin. All these interactions continue today; the life of a polar bear is dynamic, not static.

A second theme is that each polar bear is an individual. A solitary predator in an extreme environment like the Arctic must live by its wits. Thus the behavior of one bear may or may not be representative of the behavior of all bears. A single response will not answer all questions. Conditions for hunting or other environmental factors may change quickly. Consequently, polar bears are highly inquisitive. They often contemplate a situation before they act and they learn quickly from new experiences. It is also clear they remember a lot, although we do not know how much. As a result, each bear is unique because of its individual combination of experiences and knowledge.

At the risk of seeming too professorial, I have tried to supply enough background for readers to understand the polar bear in relation to its environment and how it got there. I made a list of the kinds of questions people asked so as to include the answers in the text. The chapters were laid out to give complete coverage of the different aspects of the biology and conservation of this magnificent animal. Because so many students have asked me questions about how to study a polar bear, there is a separate chapter devoted to that topic as well. I wrote the book as a non-technical reference on polar bears for individual naturalists, and places like school libraries, because there has been no book available on the life and biology of this fascinating animal.

As illustrations, Dan's photographs provide an exceptional portrait of a spectacular predator. But pictures were also selected to illustrate aspects of the biology of the animals, to increase the educational value of the book.

For ease of reading, I have not put all the references into the text as one might in a scientific paper. However, the source material is given in some detail in the Bibliography for those who may wish to go further in some areas. Because science operates on the metric system, numerical values are stated metrically, with English equivalents in parentheses.

Polar bears, the seals they eat, and the arctic marine environment in general have dominated most of my working life. As our knowledge of their behavior and ecology has grown each year, so has my continuing fascination with them. Even so, there is still so much new to learn that in many ways, each additional trip is as revealing and exciting as the first. Yet for all that, there is something special about the polar bear that takes me beyond science and objective description. Sometimes when I watch polar bears, I am almost overwhelmed by a feeling of privilege because of having been able to spend so many years working with such a magnificent animal in such a spectacular environment. Through Dan's photographs and the observations that follow, I hope we will be able to share our aesthetic and scientific appreciation of polar bears.

# The First Polar Bears

Where and when did the polar bear first become a distinctly different species? We don't know for certain because the fossil remains are so rare. Most of the handful of confirmed specimens of polar bear fossils come from Europe (Scandinavia, Germany, and England), perhaps because more excavating has been done there. Undoubtedly additional specimens are waiting to be found somewhere in the vast hinterland of northern Siberia, but their chance of discovery is low.

The paucity of polar bear fossils is not surprising. The behavior and ecology of ancestral animals were probably similar to those of their present-day descendants. They would have been distributed over vast areas of sea ice at low densities and their total numbers were likely never very large. As animals died, their remains would have sunk to the bottom of the ocean or been left on the surface of the ground near the coastline. As a result, probably a fairly small number of individual animals became fossilized in the first place, so that the chance of finding their remains scattered somewhere across northern Europe and Asia is low. In addition, the polar bear developed fairly recently as a species. Consequently, compared to animals with a longer evolutionary history, it has had less time to leave fossils about to be discovered. All things considered, it is rather surprising that any fossils of polar bears have

been discovered at all. Even the seals that lived in the same habitat for a much longer period of time, and were far more abundant, are scarcely known from the fossil record.

## Pleistocene Ancestors

Most of what we know today about the origins of the polar bear is the result of years of painstaking examination of the few existing specimens by the Finnish paleontologist Bjorn Kurten (whose first name means bear). His studies trace the evolution of the bears from the geologic time period known as the Pleistocene, which extends from the present day back for a million years or so. The Pleistocene epoch fascinates modern zoologists because the earlier part of it was characterized by so many species that were much larger than their present-day descendents. It was the era of the woolly mammoths, saber-toothed tigers, the Irish elk, giant beavers, and many other species of large mammals that have since disappeared, leaving behind their smaller relatives.

The bears of that time were also larger and must have been impressive to see. In Europe, the cave bear became one of the best known carnivores of the Pleistocene because of its convenient habit of dying in large numbers in caves where the bones were well preserved. There is a particularly famous cave in Austria that contains bones from an estimated 30,000 individual bears! (If Pleistocene polar bears were considerate enough to die in caves as well, instead of out on the sea ice, we would know a great deal more about them.) North America had a large Pleistocene cave bear that lived in Florida. Another North American native was the giant short-faced bear, which had exceptionally long legs for a bear and must have been a very fast-moving animal. In fact, Kurten describes it as the most powerful predator of the Pleistocene.

Roughly a million years ago, the cave bears gave rise to a similar-looking line, the brown (or grizzly) bears that have persevered to the present day in Europe, Asia, and North America. Unfortunately, the available fossil material from the early polar bear does not make it clear exactly which species it might have evolved from. However, an analysis of characteristics of the teeth in particular suggests it developed quite recently from a brown-bear-like form. The oldest known polar bear fossil, a bone from the lower front leg, was collected near the Kew Bridge in London. It is less than 100,000 years old, which is hardly a hiccup in the evolutionary time frame. From an analysis of the bone, Kurten determined that the Pleistocene polar bear, like other mammals of the time, was much larger than the present-day form.

## The Nearest Relative

It has been suggested that the polar bear may have separated from the brown bear line somewhere on the arctic coast of Siberia. Wherever it was, the ecological setting must have been something like the present-day situation along the Tuktoyaktuk Peninsula and Smoking Hills in the Northwest Territories of Canada. There, the present-day barren-ground grizzlies reach the arctic coast in the course of their normal travels each spring and summer. There have been a number of sightings of brown bears feeding on seal carcasses out on the sea ice, although it is unknown if they killed the seal themselves or were just scavenging. One large male I saw was about 60 kilometers (35 mi.) out on the ice, far from sight of the nearest land. Inuit (Eskimo) hunters from Tuktoyaktuk have told me that they occasionally see grizzly bears hunting for seals in their birth lairs near the mouth of Liverpool Bay. I would love to watch that myself, for I think it would give one the sense of watching the polar bear evolve.

It would be possible today, if not necessarily very likely, to see both species at the same time in a number of places along the coast in that area. In such circumstances, it does not seem too far-fetched for a grizzly to start becoming a polar bear and filling the rich but vacant niche as the supreme predator of the sea ice.

As the polar bear evolved, its appearance quickly became quite different from its brown bear ancestors. To protect it from the cold and to conserve body heat, hair soon covered the entire animal, except for its nose and the pads of its feet. The coat became white to blend in with the snow and ice that became its most common backdrop. Actually, the color can vary from pure white to creamy yellow, or even light brown, depending on the time of year and the angle of the sunlight. In overall body build, the polar bear became more elongated and its skull and head grew longer. Instead of the rather dished nose and face of the brown bear, the face of polar bear males in particular develops more of a Roman nose. Polar bear cheek teeth became smaller but more jagged as an adaptation to life as a carnivore. In comparison, the cheek teeth of the grizzly are larger but flatter and smoother for grinding the vegetation that forms the bulk of its diet. As a further adaptation to the life of a carnivore, the canines of the polar bear have become larger and sharper for tearing seals apart (p. 174). Their claws are shorter and more solid than the long curved claws that are the trademark of a grizzly bear footprint in the mud (p. 39). The stockier claw of the polar bear (p. 39) is probably less likely to bend or break under strain when its possessor is running on ice or climbing steep banks.

## How Different Are Polar Bears?

Polar bears and brown bears are both species of the same genus of carnivores, *Ursus*, which is Latin for bear. The brown bear is *Ursus arctos*, which is a little repetitious since *arctos* also means bear, but this time in Greek. The polar bear is *Ursus maritimus*, which translates as the sea bear. Taxonomists recognize the close relationship of the two species by placing them in the same genus, despite the distinct appearance of both.

So far, so good. However, this is where it starts to get interesting, since one of the basic definitions of species is that they are reproductively isolated from each other. For example, a horse and a donkey may mate but the resulting offspring, the mule, is sterile.

Interbreeding between polar bears and brown bears in captivity has occurred several times but, unlike the mule, the female offspring of the first generation, known technically as the $F_1$, have been fertile in backcrosses to either parental species. Male offspring may also be fertile but apparently the evidence for that is less conclusive, probably because the experimental crossbreeding hasn't been done. The hybrid cubs tend to be white at birth but become a bluish brown or yellowish white as they get older. One was half white and half gray-brown. In backcrosses between female hybrids and male parents, some young are white and some are dark. In body contour and the length of the head, the hybrids resemble the shape of the polar bear. One hybrid in the Hellabrun Zoo in Munich, West Germany, was apparently a much better swimmer than his

brown bear mother, but not as good as his polar bear father. The apparent ease with which interbreeding takes place between these two species, and the high degree of fertility of the offspring, indicates that polar bears and brown (grizzly) bears have not become widely separated in the genetic sense.

In a related piece of research, the Norwegian biologist Thor Larsen studied some of the genetic characteristics of the proteins and enzymes in blood samples collected from 460 polar bears all over the Arctic. Unlike the differences found in other species of mammals in similar studies, in the end he was unable to identify significant variations between populations from different areas.

For subpopulations of any species to develop significant genetic differences requires a fair amount of geographic isolation and the turnover of many generations. Under these conditions, mutations, or other naturally occurring inheritable variations, become embedded in that population and help to make it unique. In a slow-breeding animal with a long generation time like a polar bear, this takes much longer than it would, for example, in mice. The low degree of genetic variability in the polar bear, along with the demonstrated ease of interbreeding with brown bears, supports Kurten's belief that the polar bear is a very young species in an evolutionary sense. It has not yet had time to develop genetic discreteness in different populations.

## The Present-Day Polar Bears

Although the modern polar bears are smaller than their Pleistocene ancestors, they are still the largest nonaquatic carnivores presently alive anywhere in the world. Adult males may weigh from about 350 to over 650 kilograms (770–1,500+ lb.). Females normally weigh 150–250 kg (330–550 lb.) although superfat pregnant individuals occasionally approach 500 kg (1,100 lb.) (p. 175).

You may read of a graded series of changes in the size of polar bears across the Arctic. Bears are said to be smaller in the area of Svalbard and become larger as one progresses west to the Bering Sea. If this is true, presumably the bears get smaller again as one proceeds from the Bering Sea back to Svalbard again. However, it is not clear now whether that cline was real or a statistical artifact resulting from the way the samples were collected from each area. For example, many of the skulls that were measured from Svalbard were collected at a time when the population was being heavily overharvested so the bears may not have reached their full growth potential before being killed. In the Bering Sea, a large number of skulls were collected from bears that were being killed far offshore in a highly selective hunt that used small planes to find the biggest trophy bears to kill. The tendency would be for these bears to be larger. The principal weakness of the earlier studies on geographic variation in size is that they were done before accurate methods of age determination were developed for polar bears. Thus, comparisons could only be made between pooled samples of adults of a variety of ages. The skulls of adults continue to grow very slowly for several years after they reach maturity. Thus, for a proper comparison of the size of animals in different areas, measurements taken from bears of the same age would need to be compared for each area before the results could be interpreted with confidence.

In any case, it always seemed to me there were some notable anomalies, or differences, that were not easily explained. For example, it is common for male bears of western Hudson Bay and Southampton Island to weigh in excess of 650 kg (1,430 lb.), which is quite comparable to the weights of Bering Sea bears. However, if significant geographic variation in the size of polar bears does occur, I think it is much more likely to be a reflection of the overall biological productivity of the marine ecosystems they live in. If this is so, some of the most productive areas, such as the Bering Sea, or maybe Baffin Bay or Hudson Bay, may also have the largest bears.

Polar bears have 42 teeth and the dental formula is written as 3/3, 1/1, 4/4, and 2/3. This refers to the number of incisors, canines, premolars, and molars respectively. The first number of each pair indicates how many teeth of each type there are on each side of the upper jaw and the second gives the same information for the lower jaw.

In the size of a polar bear's tail and ears, Allen's Rule applies. It says that the farther north an animal is found, the smaller its extremities, because of the need to conserve energy. Because they have less surface area, smaller surface areas lose less heat than large ones. For example, desert foxes have huge ears while those of the arctic fox are tiny. Similarly, it always seems to me that the black bears of the southern United States have large ears like sails when compared to the small ears of the polar bears. Likewise, the legs of the polar bear appear larger and stockier overall than those of their grizzly cousins to the south.

In seeming contradiction of Allen's Rule, polar bears have huge feet. This is probably an adaptation to two major selective pressures in their existence in the polar ice pack. First, they swim a lot; large oarlike feet help with propulsion. Second, while searching for seals just after freeze-up or when traveling from one area to another in early winter, they must often cross areas of thin newly formed ice. Their huge paws function like snowshoes to spread out their weight and keep them from breaking through in places where a human could not walk. Sometimes if the ice is really thin, they will get down on their elbows and knees to avoid breaking through (p. 69).

Another interesting adaptation of polar bear feet to life on the arctic sea ice was discovered in an unusual way. Two English doctors, Derek

Manning and John Cooper, were involved in studying industrial accidents in Britain (over a million a year) caused by people slipping. They wondered why a polar bear didn't slip on the ice and asked if I had any ideas. I got some footpads from an animal killed by a hunter and sent them over for microscopic examination. The doctors also examined the footpads of two tranquilized bears. It turned out that the pads of a polar bear's foot are covered with small soft papillae, which increase friction between the foot and the ice. There are also small depressions in the sole, but their role is less clear. It is possible that the depressions function as little suction cups and briefly increase the grip of the bear's paw on the ice when it runs, but for the moment that is only speculation. It is fascinating to think that studies of the structure of the paw of the polar bear might influence the design of safer soles for shoes worn in the workplace.

To nurse their young, female polar bears have four functional nipples. Mothers breast-feed their young for up to two and a half years on milk that is higher in energy than that of any other species of bear. The average fat content in seven milk samples examined by the physiologist Robert Jenness and his colleagues was 33.1 percent, with variations ranging from 23.8 percent to 48.4 percent. In comparison, the mean values for the fat content of brown and black bear milk was in the range of only 22 to 24 percent. Only one sample from the small (45 kg or 100 lb.) tropical Malayan sun bear was analyzed but it had a fat content of just 11 percent. The richness of polar bear milk permits mothers to raise their young in a cold environment where energy is in high demand, hence the marked contrast to the milk from the sun bear. The high fat content may be important to polar bear cubs for another reason. Unlike brown bear cubs, after they leave the shelter of the dens they are born in, they do not escape the cold of the following winters by hibernating.

They remain active and consequently have a higher energy demand during that period.

The rich milk of the female polar bear is comparable to that of aquatic marine mammals. The young of whales and walruses, for example, have very high demands for energy because they grow rapidly and must quickly lay down a fat layer of their own for insulation when swimming in cold water. Milk samples from blue, fin, sperm, and white whales all have a fat content in the range of 30 to 37 percent, which is quite comparable to that of the polar bear. The fat content of elephant seal milk is similarly high, and gets even richer as the pup gets older, ranging from about 15 to 55 percent.

## Why a Marine Bear at All?

Bears first became recognizable as bears about 20 million years ago in the geological period known as the Miocene. These first bears were tiny, about the size of a fox terrier. They were confined to land, where all bear species but one continue to reside today. With the passage of time, bears evolved into much larger animals, but not until 100,000 to 200,000 years ago did the first polar bears abandon the land for the sea ice. Consequently, the ecological problems bears had to solve as they dispersed over all the continents but Australia and Antarctica were those of land animals. Although the evolutionary origin of the bears lies with the carnivores, they became more vegetarian in their food habits. Except for scavenging, they preferred plants as food. Kurten has suggested that the large Pleistocene cave bears of Europe and southeastern North America were specialized vegetarians. Their large size and strength came about more for protection than for predation. An exception was probably the giant short-faced bear (*Arctodus*). With his long legs, he seems to have become a superpredator. In the present day, a few bears like the pandas of China, the sun bears of southeast Asia, or the spectacled bear of South America are almost entirely vegetarians. Other living species, including the brown and black bears, are predominantly plant eaters but some populations, and some individuals within populations, may feed on salmon, moose or caribou calves, or even livestock to some degree, at particular seasons.

During the winter in the more northerly areas and high in the mountains, the plants stop growing and those close to the ground become covered with snow. Some of the potential prey species, such as ground squirrels and marmots, hibernate in inaccessible burrows deep in the frozen ground. Although the bears are admirably equipped for killing animals at close range and tearing them apart, their lumbering gait is not suited to chasing prey. In most areas, animals large enough to be worth killing, such as deer, moose, or caribou, are too wary and fleet-footed to allow a bear close enough for a successful attack. Consequently, the predominantly plant-eating bears of the northern and high mountain areas had to evolve the ability to hibernate through the winter.

Why then, if the bears were all so well adapted to being land animals, did a marine bear suddenly evolve? The expression "Nature abhors a vacuum" offers a clue.

During the Pleistocene, much of the Northern Hemisphere was heavily glaciated several times and to varying degrees. During interglacial periods, while the glaciers were in retreat, the weather warmed up and the flora and fauna flourished. For example, the best-known cave bear of Europe, *Ursus spelaeus*, was abundant about 300,000 years ago in a warm period known as the Holsteinian interglacial. That mild era, in which many species of mammals prospered, was followed by the Saalian glaciation, during which more of the land surface of Europe was covered

than ever before. By the time the next interglacial period, the Eemian, had arrived about 125,000 years ago, we had polar bears.

We can speculate that as Eurasia became colder and ice began to cover the Arctic Ocean, a new year-round source of especially rich food became accessible to bears along the coast: seals. Probably few, if any, other land animals hunted seals much during the interglacial period so they may not even have learned to be wary. Maybe at a place like the Tuktoyaktuk Peninsula in the Northwest Territories of Canada, ancestral brown bears first encountered seals on the landfast sea ice. Possibly they scavenged on dead ones initially before they discovered they could catch live seals at cracks in the ice simply by sitting still and waiting for them to come up to breathe. Bears learn quickly, so it is easy to imagine that if an individual accidentally learned how to catch a seal, the reward would be strong enough that the practice would soon be repeated. Other animals would come to scavenge and quickly learn to hunt by imitation. Cubs would be taught the trick by their mothers. Suddenly there was a new way to make a living, an opportunity ecologists call a niche. Most important, it was an unoccupied niche, with no other animal taking advantage of it.

At this point, natural selection probably began to operate intensely and quickly, favoring the survival of animals with traits that enabled them to capture seals. Among brown bears today, for example, there is a wide range of natural coat colors from dark to light. It is reasonable to suggest that lighter colored bears were less visible on the ice. Because of this, they were more successful at capturing seals. That enabled them to feed more frequently, survive better, and reproduce more often. Successive generations over time produced the white coat that has become the trademark of the polar bear.

Initially, the seal-hunting habit might have been

localized, in much the same way that brown bears living in a few river valleys in Alaska have learned to depend upon spawning salmon for food. However, seals would have been present thoughout the northern seas. Consequently, there was ample opportunity for this behavior to spread to bears in other areas. As the ice sheets of the Saalian glaciation marched south into Eurasia, the advancing front would have pushed the brown bears away from the arctic coast, separating them forever from their white cousins, who were equipped by then to stay behind. In isolation, the new white-coated "brown" bears would evolve even more rapidly. By the time of the Eemian interglacial, about 125,000 years ago (and likely before that), the polar bear was already well developed as a separate species, as the 100,000-year-old specimen from London indicates. By that time, the distribution of the polar bear had probably become circumpolar as well. Even as the northern coastline of the continents warmed in the coming interglacial, the polar bear could remain on the pack ice of the arctic basin hunting seals and denning on the most northerly islands or on the ice itself. Thus, even though polar and brown bears mated at about the same time of year and would still be capable of interbreeding, their habitat requirements would continue to keep them apart. The same conditions apply along the Tuktoyaktuk Peninsula today. Left alone for another 50,000 or 100,000 years, polar bears will probably continue to evolve to the point where they can no longer interbreed with brown bears. Then they will be a true species in every sense of the genetic definition.

## Adapting to the New Life

As it evolved, the ice bear had to cope with ecological factors different from those confronting land bears. First, so long as the bear could remain on the ice, it could catch seals all year round. Consequently, there was no longer a need to hibernate during winter in order to survive the food shortage that occurs at that time for brown bears. Pregnant females still went into dens during the winter because the tiny young required shelter. They needed time to grow big enough to withstand the cold and follow their mothers as they hunted on the open sea ice.

Although the present-day polar bear is smaller than its Pleistocene ancestors, it is still the largest of the modern nonaquatic carnivores. As usual, there is a good reason. A larger body is a more efficient conserver of energy than a smaller one.

Meanwhile, the seal population was not standing still. Under the pressure of predation by polar bears, seals began to develop avoidance strategies. Over thousands of years of attack or escape, living or dying, bears and seals shaped each other's ecology and behavior, habitat use, reproduction, and population dynamics. In the process, the polar bear has become a true arctic marine mammal, in the sense that it now depends entirely on the sea for its existence. Throughout this book, the theme will be to describe how the polar bear has evolved to meet the ecological requirements of its unique life and how it is now coping with its most recent and potentially stressful environmental strain: modern man.

# The Original Polar Bear Watchers

The first polar bear watchers probably shared origins similar to those of the polar bears themselves. The boreal forests of North America and northern Eurasia afforded an ecological niche to human populations for a few tens of thousands of years. However, until quite recently, there were no people living on the tundra or the frozen rim of the Arctic Ocean. To the forest dweller, it was unfamiliar. The climate was harsh and the secrets of survival there were not easily learned.

Finally, about 4,000 years ago, like the polar bears before them, the Paleoeskimos dared to abandon a land-based existence to begin a new culture supported by the polar seas. Where these people came from is unclear, but the structure of tools found in their middens in Alaska suggests that the Paleoeskimos came from northeastern Siberia. They probably crossed to North America on the frozen ice of the Bering Sea. Possibly they came in boats, although there is no evidence of it. Most dramatic, though, is that radiocarbon dating of artifacts collected all across the Arctic gives a very similar age. This indicates that once the Paleoeskimos figured out how to exploit this unoccupied niche as human predators, they spread extremely rapidly from Alaska to Greenland in much the same way as polar bears must have done some 200,000 years earlier.

A major cultural change occurred about 2,500 to 3,000 years ago when the Dorset culture appeared. The first efficient harpoon heads and the skulls of dogs are found in their sites, indicating major improvements in the development of their proficiency at both hunting and traveling. Here we also find the first unmistakable signs of the importance of polar bears. Polar bear watching had begun in earnest.

To the Dorset people, polar bears were of major importance. Their campsites yield small carvings of all the animals they hunted. They sculpted walruses, seals, caribou, fish, and birds, but most often, they carved polar bears. The Dorset carvings of bears show an especially high degree of detail of the skeleton and joints. Some carvings are so stylized they can only be identified by the characteristic skeletal markings. They also made bear masks and miniatures of some of the hunting weapons. Archaeologists interpret the predominance of carvings of this, the most powerful and dangerous of the arctic mammals, to mean that the carvings represent the spirits of polar bears rather than bears themselves.

All evidence considered, it is clear that the polar bear played an important role in the religious practices and ceremonies of the time. Unfortunately little more is known about the Dorset people. By the time the first Viking explorers arrived in southwestern Greenland a thousand years ago, only the ruins of their dwellings remained. What they believed and knew about polar bears we do not know. Nevertheless, it is clear that the great white bear was valued and respected.

As the Dorset people were disappearing from the eastern Canadian Arctic and Greenland, possibly as a result of changes in the climate, the Thule culture had developed on the northern coast of Alaska. The Thule people were the supreme maritime hunters of the Arctic. So advanced were their skills that they even developed techniques for hunting the great bowhead whales. They quickly spread across the Arctic, reaching northwest Greenland by about A.D. 1200. The remains of their houses are strewn with the bones of all the marine mammals of the Arctic, including the skulls of polar bears.

Much of the knowledge and tradition of the Thule people has been passed down to their descendants, the present-day Inuit. Through the journals of the explorers, traders, missionaries, and, thank goodness, scientists like Boas and Rasmussen, we have some records of the relationship between the early Inuit and polar bears. The Inuit have a very strong tradition of oral history through which they preserve the past. Since contact with European civilization has been so recent, in some areas, almost within living memory, anthropologists and others are still recording oral history before it is lost forever. Incredibly, there are still a few people in the Arctic who have gone from the Stone Age to the Space Age in a single lifetime.

## Use of Polar Bears by Inuit

The Inuit killed polar bears for clothing, sleeping skins, and food for themselves and their dogs. There were also a number of specialized but local uses. For example, a hunter waiting for a seal to come up to its breathing hole in the bitter cold of winter might sit on a piece of polar bear skin on a snow block to make a warm, dry seat. He might rest his feet on another to help keep the cold from penetrating.

In the Ungava area of northern Quebec, apparently hunters would use a pad of polar bear skin to protect the elbows when crawling up on seals. Inuit commonly used a piece of polar bear skin as a brush to spread water on sled runners in order to coat them with ice in the cold weather. The bear's canine teeth were often kept for making ornaments or amulets.

Bears were sometimes killed while hunters were out looking for other animals such as seals. But the Inuit also specifically hunted bears on the ice in March and April. Most commonly, hunters would follow a fresh track with a dog team. When they neared the bear, they would unleash the dogs to let them rush at the bear and distract it by barking and snapping at its heels. This gave the hunter a chance to close in and shoot the bear with arrows, or impale it with a lance.

Dogs were used to sniff out female bears with cubs in their maternity dens. The hunters would then kill the bear by plunging a spear through holes in the snow. Hides of the cubs were particularly sought after and the meat was regarded as a special delicacy.

Of all the Inuit, those of the High Arctic and Greenland seem to have valued polar bears most highly. In those areas, they formed a greater proportion of the game taken in a year than they did for groups elsewhere who depended mainly on caribou and seals. The so-called Polar Eskimos of northwestern Greenland hunted and used polar bears more than any other group. They also appear to have developed the most elaborate rituals involving bears. To this day, the wearing of polar bear pants by a boy or a man is a sign of status.

## Inuit Spiritual Beliefs and Practices Concerning Polar Bears

As important as polar bears were for food and other materials, they were not the mainstay of life as were seals or caribou. Consequently, we must go to the cultural and spiritual aspects to gain a sense of the full traditional significance of polar bears to the first polar bear watchers.

The prehistoric Inuit had a dual relation with most animals. They knew how to kill and use their bodies but when they did so, they believed they had to defer to their spirits. This required that they act according to quite specific beliefs and rituals or the animals would be offended and withhold themselves from the hunters. An Igloolik Inuk hunter, Ivaluardjuk, explained the essence of these beliefs to the famous Danish ethnologist Knud Rasmussen as follows: "The greatest peril of life lies in the fact that human food consists entirely of souls. All the creatures that we have to kill and eat, all those that we have to strike down and destroy to make clothes for ourselves, have souls, like we have, souls that do not perish with the body, and which must therefore be propitiated lest they should revenge themselves on us for taking away their bodies."

Beliefs and practices varied between areas but there were some common themes. One was that the spirit of an animal might be chosen to be the *tornaq* (spiritual guardian) of a particular individual. With the exception of Sedna, the legendary goddess of the sea who was generally held to be the supreme being, the most powerful spirit of all was that of the polar bear. Consequently, the shaman usually had the polar bear as his *tornaq*. It is certainly no coincidence that when the mighty *angakoq* (shaman) of ancient Greenlandic legend made his daring and adventure-filled trip to the moon and back, he was accompanied by his polar bear *tornaq*.

Another important concept was that the spirits of men and bears were interchangeable. To start with, bears were the most powerful and dangerous of all the animals, so that the killing of one was a major event. More significant, as pointed out by the American anthropologist Irving Hallowell, is the fact that bears have so many "human" traits. They can stand up and walk on their hind legs, or sit and lean against something as if resting and thinking. They also eat many of the same foods, both plant and animal, that people do. However, of greatest importance is probably the fact that the musculature of a bear's skinned carcass has an uncanny similarity to the body of a man. This undoubtedly explains why, in so many Inuit legends, when the polar bears are inside their own houses they are people; they put on their hides when they go outside to be bears again. There are parallels in legends about other species of bears throughout the Northern Hemisphere.

The Inuit believed that a bear would give itself to a hunter only if it were treated properly after death. Thus, hunters were careful to observe various ceremonies after killing a bear. The most widespread practice was to observe a strict taboo against hunting polar bears after killing one to allow sufficient time for its soul to return to its family. Rasmussen recorded a legend from the Netsilik Inuit ("people of the seal") of the Central Canadian Arctic that outlines this belief. In it, a woman had unwittingly entered an igloo inhabited by a family of polar bears. She became frightened and hid behind some sealskins that were hanging on the inside walls. As she listened to their conversation she learned how the youngest bear had gained respect for the Inuit, who he had once thought were just figures of skin and bone. This bear had been out hunting humans and had been killed by a man who later gave him a death taboo and several wonderful presents. These actions set the bear's soul free and he was able to return to his family again after four days.

Later, when the woman escaped, she returned to her village and told the people what she had learned. Because of this legend, the Netsilik, Copper, and Inland Inuit observed a strict taboo on hunting for several days after killing a bear. Curiously, this abstinence lasted for five days after killing a female bear and four days for a male.

In the Hudson Bay and southeastern Baffin Island areas, the nonhunting period during which the polar bear's spirit was thought to linger for three days. This was also identical to the length of time believed to be taken by a human spirit in those areas to leave its body and return to its ancestors.

The Copper Eskimos had another practice linking human beings to bears. If a male bear was killed, the hunter would give it a miniature bow and arrow, while a female bear was given a needle holder because, like humans, the man needs his hunting weapons and the woman her domestic tools.

In rituals, only parts of a bear usually were used to represent the the entire animal. The skull was most important to the Asiatic Eskimos, Greenlanders, Polar Eskimos, and those in the Central Canadian Arctic. To the Polar Eskimos, as well as those of inland Alaska, the MacKenzie Delta, and the Central Canadian Arctic, the skin was important. Parts of the animals' intestines were also important to the Inuit of Baffin Island, the Chukchi Sea, and the Igloolik areas.

In parts of Greenland, before contact with Europeans, some of the practices were apparently quite elaborate. The Scandinavian anthropologist Helge Larsen reported that the Polar Eskimos regarded the soul of the polar bear as more dangerous than that of all other animals. Consequently, posthunting rituals had to be carried out especially carefully. Particular attention also focused on the head of brown and black bears in boreal areas of Asia and North America. As an offering to the soul of a male polar bear, a harpoon was hung over its snout for five days, along with an offering of blubber and meat. A female bear would be given a piece of sealskin as well.

In southern Greenland, a hunter placed the head of the bear on the lamp platform facing southeast, which was the direction the bears came from in that district. He covered the eyes and blocked off the nostrils with moss or other materials so the bear's soul could not see or smell him. Fat was smeared on the jaws and articles such as boot soles, knives, or beads were placed on top of the skull. The fat on the mouth was to appease the bear because they were known to like greasy food. The materials on the head were for the ancestors of the hunters who, they believed, had sent the bear to collect them. Like their kin to the south, the Polar Eskimos also refrained from hunting polar bears for five days after a kill in order to allow sufficient time for the bear's soul to return to its home.

The American anthropologist Charles Hughes reported that in the days before the acquisition of firearms at St. Lawrence Island in Alaska, bears were killed with lances. The hunter who accomplished this dangerous feat was honored by five days of ceremonies similar to those judged appropriate for a whale. The head of the bear was put in a corner of the room with its mouth open and decorated with beads and so on appropriate to its sex. All hunting ceased while stories were told and songs were sung. After five days, the skull was boiled and the pieces of flesh that became dislodged were either thrown into the air for the spirits or placed in the fire to appease their forefathers. Afterward, the hunter put the skull on the graves of the clan's ancestors, along with other bear skulls. St. Lawrence Islanders also believed that a hunter who wounded a bear must track it down and kill it so as to release its soul. If this was not done, the animal's soul would be deeply offended and cause sickness and harm to the hunter.

## A Polar Bear Legend

There are many Inuit legends and tales about polar bears. One of my favorites, *The Great Bear*,
was recorded by Rasmussen (1921, p. 81) on the
coast of West Greenland shortly after the turn of
the century. It goes as follows.

A woman ran away from her home because her
child had died. On her way she came to a house.
In the passage way there lay the skins of bears.
And she went in. And now it was revealed that
the people who lived in there were bears in human
form.

Yet for all that she stayed with them. One big
bear used to go out hunting to find food for them.
It would put on its skin and go out, sometimes
staying away for a long time, and always return
with some catch or other. But one day, the woman
who had run away began to feel homesick, and
greatly desired to see her kin. And then the bear
spoke to her thus:

"Do not speak of us when you return to men," it
said. For it was afraid lest its two cubs might be
killed by the men.

Then the woman went home and there she felt a
great desire to tell what she had seen. And one
day, as she sat with her husband in the house, she
said to him:

"I have seen bears."

And now many sledges drove out, and when the
bear saw them coming towards its house, it felt so
sorry for its cubs that it bit them to death, that
they might not fall in to the hands of men.

But then it dashed out to find the woman who
had betrayed it and broke into her house and bit
her to death. But when it came out, the dogs
closed round it and fell upon it. The bear struck
out at them, but suddenly all of them became
wonderfully bright, and rose up to the sky in the
form of stars. And it is these which we call Qilug-
tussat, the stars which look like barking dogs
about a bear.

Since then, men have learned to beware of bears,
for they hear what men say.

I particularly like that story because it contains
so much of the essence of the mythological relationships between the Inuit and the polar bears
before the arrival of Europeans. There is the
sense of the bear being a particularly powerful
spirit, able to turn dogs into stars. There is the
recurrent theme from many legends that the spirits of men and bears are interchangeable. Finally,
there is the lesson of how one behaves toward
bears and talks about them.

Talking respectfully about polar bears is still
deeply ingrained in many Inuk hunters today, especially older and more traditional individuals.
For example, just recently an exceptionally good
hunter who is a friend of many years confided to
me that two young fellows who were out hunting
polar bears would not succeed. "They been brag-
ging too much about getting the bear. The bear
won't let himself be killed by someone who talks
like that." I made no comment because the two
young hunters were in an area where bears are
usually abundant and I thought it quite likely
they would each get one. Consequently, I was a
little surprised when they came back empty-
handed. My friend on the other hand, saw nothing unusual and made no further comment on
the matter. I like the deep respect that older
hunters have for polar bears, for I feel the same
way myself.

## Polar Bears and the Nenets of Northwestern Siberia

The Nenets people, who live along the arctic
coast of the Soviet Union, from the White Sea to
northwestern Siberia, also gave polar bears a
special place in their traditional beliefs. They
erected altars of bones from several species in
noticeable places. In particular, the skulls of polar bears were especially featured. Savva Uspenski,
the Soviet polar bear biologist, measured one

such ceremonial mound of reindeer and polar bear skulls that was about 5 meters in diameter and 3 meters high. Another had fifty-five polar bear skulls in it. In northern Yamal, Uspenski was given permission by the Samoyeds to remove some skulls for scientific studies but they asked him not to take any of the fresh ones because it might affect the success of next year's hunt.

The Nenet people also practiced other rituals in earlier times, such as the taking of an oath while holding the muzzle of a skull or a paw. In some regions, women were banned from eating polar bear flesh. Canine teeth were used as amulets against visits by deceased persons and attacks by brown bears.

## The Age of the Explorers

From the sixteenth century on, European explorers ventured into the Arctic with increasing frequency in search of the fabled Northwest Passage as a short cut to the far east. Success at finding new lands (if not a seaway), mapping them, and developing hydrographic charts was almost a guarantee of wealth and military promotion. Rewards were posted for the successful explorer, and in 1745 cash payments were even provided for by an Act of Parliament in Britain. To help maintain interest, consolation prizes were offered for the attainment of lesser goals in the quest.

With such inducements, the 1800s became the golden age of arctic exploration. Unfortunately, some expeditions ended in tragedy. After spending the winter of 1845–46 at Beechey Island, Sir John Franklin set sail with 129 men in the *Ebebus* and the *Terror*. Franklin's objective was to discover the Northwest Passage, but he and his crew disappeared into the arctic mists without a trace. For years, expeditions used the mystery of Franklin's disappearance as a justification for raising funds to support investigations to al-

most every corner of the Arctic, allegedly in search of Franklin. It was no secret that the real goal of most of the explorers was the discovery of the Northwest Passage. The result was a tremendous amount of new geographical information but, in general, the advances in biological knowledge were rather limited. Most voyages to the Arctic did not include naturalists of the caliber of Hooker, Banks, or Darwin, who accompanied similar voyages to the southern oceans. Knowledge took second place to the quest for money and fame. Many irreplaceable scientific opportunities were lost as a result.

To the explorers who found a bleak and inhospitable environment awaiting them, the polar bear loomed larger than life and was much feared. The journals of almost every expedition contain similar accounts of bears approaching the ship, being hunted on the ice, stalking people, and being shot. The deaths of countless numbers of bears are chronicled in generally theatrical and repetitious fashion. Unfortunately, the reports are generally devoid of useful information.

## The Transition Period for Arctic Science

Unlike the Inuit, the nineteenth-century explorers were largely insensitive to the world around them. They came to conquer the Arctic, not to be a part of it. Fortunately, with the turn of the century, a new generation of scientific explorers appeared. Men such as Boas, Rasmussen, and Stefansson came to learn from the indigenous polar bear watchers and record their knowledge. They represent the transition to the new polar bear watchers (modern scientists) who have spent much of their lives using new tools and approaches to learn about the bears in different ways.

Sometimes scientists have simply found the reasons for things that the Inuit already knew. For example, polar bear liver is not safe for humans to eat. This is because the concentrations of vitamin A, up to 15,000 to 30,000 units per gram, are so high they are toxic. On the other hand, using new technologies and insights, scientists have contributed much to our factual understanding of polar bears. In that sense, the rest of the book examines the findings of the new polar bear watchers.

Black bears are only one-third the size of polar bears. Their ears are large and their legs are long and thin in proportion to their body. They often have a white patch on the chest.

Male brown bears have a massive, heavily muscled head with a dished face and a hump on the shoulders. Light color phases such as this may have given rise to the white coat of the polar bear.

Tracks of a polar bear walking away to the upper left of the picture. The hind footprint at the bottom right alternates with the front.

Typical track of a large male, dragging his feet as he walks

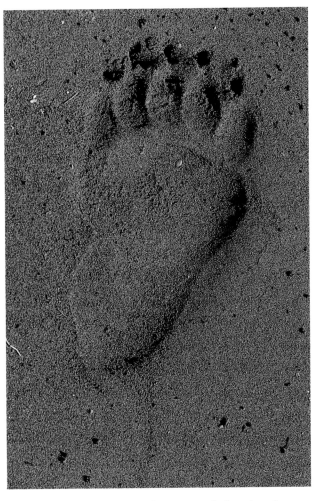

Hind footprint of a brown bear in mud showing the imprint of the characteristic long claws

Hind footprint of a polar bear in snow. No claw marks are visible.

Bottom of the front paw showing that it is completely furred

Upper side of the front paw showing the short stocky claws

Even by eight months old, the face is beginning to fill out.

The head of a polar bear cub is small and delicate. Its fur is soft and much whiter than that of older bears.

Through about two to five years of age, the head gets longer but not much wider.

At fifteen years of age, the overall appearance of the head is massive and scarred from fighting with other males over females.

The heavy muscles develop on the top of the head of adult males and the overall appearance of the head is much wider.

In old age (twenty or more years), the head is extremely scarred but the muscles begin to atrophy. The skull is still large but the appearance is no longer massive.

Typical maternity den site at the edge of a small lake inland from the coast of western Hudson Bay

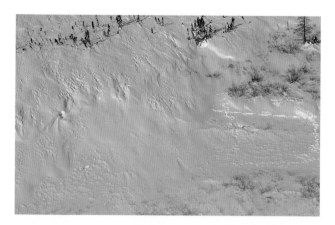

The exit hole from the maternity den is on the left. Tracks show where the female and cubs have walked about outside to acclimate. The pit on the right was used by the mother to nurse her cubs.

Looking out toward the entrance from the inside of a maternity den. (Photograph © Ian Stirling.)

A triplet litter watches what is going on around it.

Three-month-old cub nursing

Tracks of female polar bear and her two cubs leaving
the maternity den to return to the sea ice to hunt seals

Three-month-old cubs climbing onto their mother's
back for a ride while traveling through deep snow

A polar bear can sleep in any position.

A resting polar bear may lie in a snowdrift until he is completely covered over and keep on sleeping inside.

Resting in a day bed

Sleeping with a block of ice for a pillow

Resting anywhere, anytime, to conserve energy is an important part of the life of a polar bear.

# How Do You Study a Polar Bear?

Studying polar bears presents a real challenge. Imagine some of the problems.

To start with, polar bears are large and capable of doing a lot of damage to a person quickly. Working with them safely requires caution and patience. Even in places where there are lots of bears, they tend to be distributed at low densities over vast and usually remote areas. Much of the time they are offshore in drifting pack ice that is only accessible by helicopter. Sometimes you work all day and only catch three or four bears. Occasionally, you don't even see one. It is not unusual for a study to take several years to complete.

The weather can be cold, windy, or foggy. Sometimes several days may pass before the weather is good enough to fly again. In some areas, it is necessary to work at distances of 240 km (150 mi.) or more from the nearest fuel cache. Mechanical failure of aircraft is always a possibility. In not so many years, my crew alone have been involved in two helicopter crashes, two fixed-wing crashes, two helicopter engine failures, and too many narrow escapes to recall. Miraculously, no one has been hurt.

Recognizing these difficulties, you may appreciate that conducting research on polar bears is expensive, time-consuming, and potentially dangerous. On the other hand, there are few animals anywhere in the world as fascinating and rewarding to study.

Historically, studies of large carnivores worldwide fall into two categories: those done before the development of safe, fast, and economical methods of immobilization, and those done afterward. Until recently, scientists working in the wild on large predators such as bears, tigers, or leopards, have been limited to describing the behavior of unknown animals seen in the distance, deducing information from tracks and the remains of prey, or collecting specimens from dead animals. Only occasionally would the field scientists be able to recognize an animal from natural markings or scars so that its individual history could be followed. Although a great deal of important information was collected in this way, you can see its limits and the shortcomings of its interpretation.

The development of immobilizing techniques changed all this. Once a wild animal could be temporarily immobilized, it became possible to handle and tag it, collect specimens, and attach radio collars so as to track its subsequent movements. It became possible to handle animals in numbers large enough to be meaningful in studies of population dynamics, seasonal movements, and discreteness of populations. Most important of all, a researcher could capture, mark, and recapture the same animals as often as necessary. "Longitudinal studies" became possible in which individual animals could be followed for many years, even a lifetime, without interfering with their normal activities. The importance of this breakthrough to the quality of the research, and the conservation programs that have been developed as a consequence, cannot be overestimated.

## Learning to Immobilize Polar Bears for Research

Although it all seems rather routine today, there were some frustrating failures when the first drugs were tested on wild polar bears in 1967 near Barrow, Alaska.

The method works like this: After you locate a bear, you estimate its weight and load a hypodermic syringe, or dart, with the required dose of a tranquilizing drug. The amount used depends on the animal's size. You insert this projectile into a specially modified dart rifle. The helicopter pilot places you just above the fleeing animal and you shoot the dart into the heavy muscles of its rump or shoulder. On impact, the needle on the dart penetrates the hide and the drug injects automatically. If all goes well, the bear goes to sleep in five to fifteen minutes.

The first drug tried on wild polar bears was Succinylcholine chloride. It had worked well on black and brown bears that had been captured in snares, and there was no reason to think it would not be effective on polar bears as well. In the first trial, ten bears were injected. To the dismay of

the researchers, four bears died, two were unaffected, and only three were tagged—not an impressive record. Possibly the additional stress put on a polar bear by chasing it with a helicopter caused it to overreact. Use of that drug was discontinued.

Soon afterward, several biologists tried Sernylan (Phencyclidine hydrochloride). Overall, it worked quite well. The bears went to sleep quite quickly and, by watching from a safe distance, it was easy to tell just what effect the drug was having. Within a few minutes after being injected, a bear would stagger a bit and then wobble back and forth while walking ever more slowly. After standing for a moment it would lie down. In the last stages, some bears would still wave their heads after lying down, giving a false impression of alertness. Sometimes a bear would not get enough of the drug into its system. With experience you can tell from a distance whether it is safe to walk up and give a booster injection by hand or whether another dart delivered from the air might be needed.

The main disadvantage of Sernylan was that it sometimes caused convulsions. While these do not seem to harm the bear, they are unpleasant to watch. As biologists, we are all sensitive to the welfare of individual bears and disturbed when one convulses. On the plus side, the convulsions seemed to relieve built-up stress in the animal and afterward they breathed deeper and more steadily.

Another of Sernylan's disadvantages is that if the dose is too heavy, the animal might simply stop breathing. Its heart keeps on pounding away, however, so you can keep it alive by giving it artificial respiration until the drug wears off and the bear begins to breathe on its own again. And just how do you give artificial respiration to a polar bear? Mouth to mouth? Not quite. You lay the bear on its side, take a handful of fur over the rib cage, and lift. This opens the lungs like a bellows and they fill with air. Then you push down and squeeze the air out. What seemed to work best was to repeat this cycle three or four times in quick succession with intervals of 10 to 15 seconds in between.

Few bears actually required artificial respiration and most of those that did required it for only 15 or 20 minutes. However, on one occasion I had to keep it up for 2 hours and 45 minutes. It happened in early May, just as the period of 24-hour daylight was approaching. Shortly before midnight, I spotted an adult female bear and darted her. She went down quickly. On the ground, I discovered that she was much thinner than she had seemed from the air. The dosage was too strong and she stopped breathing fairly soon after we reached her, so I started artificial respiration. It was cold, about $-20°$ C ($-2$ on the Fahrenheit scale) and windy with blowing snow. I kept on pumping air into her until 2:00 A.M. I knew she would survive but I was still awfully glad to see the old girl start breathing on her own so we could go home, have a cup of tea, and go to bed. As this was the longest period of artificial respiration that we had ever given to a bear, I was concerned about the possibility of long-term effects. However, we caught her again the following spring. She was in excellent health and accompanied by newborn cubs. It was clear she had not suffered unduly from her interlude with science.

One of our regular practices in the Inuit settlements we work from is to take some of the hunters out with us to see how the polar bear work is done. As no animal is more important to the Inuit than a polar bear, it is particularly important that they understand and participate in the fieldwork. Also, the older and more experienced hunters have accumulated a wealth of information on the biology of the polar bears in their area, through years of astute observation and endless discussion.

Sometimes, though, one fails to appreciate the act of faith required for some of the older hunters in particular to accept what is happening at face value. Nowhere is this technological gap more apparent than when immobilizing polar bears. The sight of a large male bear lying on the snow, breathing perfectly normally with the eyes open and watching his movements, makes many hunters very nervous. After a lifetime of treating live polar bears as dangerous in the extreme, some of the hunters have a hard time adjusting to the idea that a chemical process they don't completely trust can keep a bear from suddenly jumping up and attacking.

This was brought home to me one cold windy day along the west coast of Banks Island about fifteen years ago. I darted a thin adult female a little too heavily and she stopped breathing. After giving artificial respiration for about 20 minutes, I paused to check for the first light breaths that indicate the beginning of recovery. However, the wind was blowing so strongly that her fur was waving about too much to be able to detect any slight movements her rib cage might be making. I placed my bare ear against her nose and cupped my hand over the top to make a better seal. This way, you can feel if there are any faint puffs of warm air from her lungs. I did this a couple of times and as I felt her beginning to breathe more strongly, I sat back on the snow to rest a bit. I looked up and noticed that my friend had backed away several yards. He was nervously shifting back and forth from foot to foot. Finally, concerned for my safety, he blurted out, "You know, if you keep on doing that . . . you're not going to live very long!"

The big shortcoming to Sernylan, though, was not its effect on bears but its effect on humans. It proved to have hallucinogenic properties. During the 1970s, it came into widespread use in the streets under the nicknames of PCP and angel dust. With drug abuse increasing, the United States banned its manufacture and sale by 1980. We had to look for a replacement.

Over the next couple of years, we tested various replacements but none was satisfactory. One combination of two drugs, Rompun (Xylazine hydrochloride) and Ketamine (Ketamine hydrochloride) worked reasonably well but had serious disadvantages. When a polar bear runs, its big well-insulated body quickly generates a lot of heat. To get rid of excess heat, a bear pants, or breathes rapidly. However, when injected with the drugs we were trying, a bear's breathing rate would drop to about 4 to 6 breaths a minute from a normal rate of 10 to 15. The heart rate would also be depressed to 10 to 20 beats a minute from a normal rate (for a drugged bear) of 55 to 70 beats per minute, depending on the size of the animal. In cool weather, this is not a serious problem. But in warm weather an animal may overheat. When drugged, the bear may not be able to increase its respiration and heart rate in order to cool itself off, and it could die. Because of this danger, we were severely restricted as to what we could do during warmer weather.

A second disadvantage to the new drugs derived from the different ways in which a drugged bear reacted to them. Often, it was difficult to tell from a distance when it was safe to approach. The bear did not wave its head about in the last stages of drug-induced slumber, as it did with Sernylan. A bear might be lying motionless but still be quite capable of standing up . . . suddenly . . . and several did! Fortunately, polar bears are wonderfully forgiving of biologists and we just ended up having our wits scared out of us. There was also the problem of safety for the bear. If it lay down in an area of muskeg or sea ice where there were small pools of water, it might get its nose in the water and drown before we could safely move it. Lastly, the stress from additional noise or handling was sometimes enough to stimulate a bear to get up without warning when the drug began to wear off.

One of my graduate students, Malcolm Ramsay (now at the University of Saskatchewan) read about an antidote to the two-drug combination,

Yohimbine (Yohimbine hydrochloride). We thought it might help overheated polar bears get up and look after themselves more quickly. The Canadian Wildlife Service veterinarian, Eric Broughton, came out to help us. We set up camp just south of Cape Churchill on the the Manitoba coast of Hudson Bay during the late summer. The camp was encircled by a high fence. Our plan was to drug a bear and bring it back to camp in a cargo net slung beneath the helicopter. With the bear outside the fence, we could beat a hasty retreat to safety, if necessary, after administering the antidote. Malcolm went out with our pilot, Steve Miller, drugged a young bear about 150 kg (300 lb.), and brought him back to the camp. Eric and I noted the animal's body temperature, heart rate, breathing rate, and general appearance. He was totally unconscious. We had no idea how fast this antidote would work so we opened the gate into the compound and cleared everything out of the way between the bear and the gate. Then we injected a tiny amount of antidote into the small vein on the underside of the tongue. We watched cautiously from within the compound. Several minutes passed and nothing happened. We carefully approached the sleeping bear with a long pole and gently prodded it . . . no response. A few harder prods and still no response. Since this was the first bear to be tested, we had no idea what to expect. Had we given too little of the antidote? Would it work on bears at all? We didn't know.

One of the first tests to use on a bear to see if it is recovering is to pull its tongue. When you do this, you hold the upper jaw just behind the nose and lift, thereby opening the mouth. Then you reach in with the fingers of your free hand, at the side of the mouth, just behind the large canine teeth. You take hold of the tongue and give a few gentle tugs. If the bear pulls back, he is beginning to wake up. The reason you reach in from the side is that there is a space on both the upper and lower jaws, between the canine teeth and the first premolars, called the diastema. There are no teeth there. Thus, if the bear does happen to close its jaws, you may get a bruise, but you won't get bitten.

I did this but got no response. I carefully laid his tongue back in his mouth so he would not bite it when I closed his jaws. A few minutes later, I went to test the tongue response again, fully expecting no reaction. This time I made the mistake of reaching into his mouth from the front, sticking my fingers between the rows of large sharp incisor teeth. Just as I was feeling for the tongue, the jaws snapped shut. His teeth pierced the sides of both my thumb and forefinger, narrowly missing the bone. With his eyes becoming ever more alert and fixed on me, he hung on. There was no doubt now that the antidote was working! Eric tried to pull the jaws apart with no success. Then he slapped the bear on its sensitive nose. He momentarily loosened his grip and I pulled my hand free. A few minutes later, the bear got up and sauntered away, no doubt pleased at having gotten his own back on the biologists.

After more testing, we found that bears could be returned to normal in a matter of minutes through use of the antidote. In fact, some got up so quickly that we began packing our equipment into the helicopter and getting ready to leave before giving the injection. Nevertheless, the other difficulties continued to concern us. In particular, we felt it cut the safety margins too thin, especially for an inexperienced biologist.

The scientific literature reported another experimental drug developed by Parke Davis, a subsidiary of the Warner Lambert Company, called CI-744. It was Telazol, a one-to-one mixture of Tiletamine hydrochloride and Zolazepam hydrochloride. From the small amount of information available, it appeared to be especially good for bears. Also, I knew that George Schaller had used Telazol to immobilize pandas in China in order to put radio collars on them. Since it would

probably have caused an international incident if a drugged panda died, I knew George would have used the safest drug he could get his hands on.

With the outstanding cooperation of the Warner Lambert Company, we tested Telazol on polar bears during both the critical hot months of summer and the very cold period of early spring. The results were nothing less than spectacular. Bears became immobilized quickly after receiving the drug. From their behavior, it was easy to tell when you could approach them safely. They held their heads up during the last stages of induction but did not have convulsions. They had a high degree of tolerance to doses that were heavier than required. They recovered more quickly than they did from other drugs, except where an antidote might be used. They were able to modify their breathing and heart rates to regulate their body temperature as required. Recovery took place in a series of well-defined stages and the bears showed no sign of aggressive behavior after they woke up. Because they went to sleep so readily and recovered so quickly, we were able to save time, which enabled us to catch more bears in a day, and thus work more cost-effectively. Telazol was clearly the best drug any of us had ever seen for polar bears and it is now being used almost exclusively.

## What Do You Do with a Drugged Polar Bear?

Once the technical aspects of drugging polar bears were worked out, the collection of data quickly became standardized. Helicopters became our routine (if expensive) transportation for searching for wild polar bears. After an immobilized bear is approached, it is first checked to ensure it is lying comfortably and able to breathe properly. Then a series of standardized procedures are undertaken. The length and axillary girth (around the body under the armpits) are measured (p. 174). If available, scales are used for weighing the bears, although the body weight can be calculated quite accurately from the axillary girth alone. White delrin tags, each with the same identifying individual number, are put into each ear. We used to use colored tags but the bears appeared to notice them and sometimes groomed them off each other. This was particularly true of mothers and cubs. The polar bears do not seem to notice the white tags. The tag number is tattooed into the inside of both upper lips (p. 174). Even if the tags are lost, we can always reidentify the bear because the tattoos are permanent (p. 174). The numbering system for the tags is coordinated between all the countries that have polar bears (Canada, Denmark, Norway, USA, and USSR) to ensure that the same number is not used twice. A small premolar tooth, about half the size of a cribbage peg, is removed from just behind the canine tooth so the age of the animal can be determined. In the laboratory, we soften the tooth in acid, slice it in very thin sections, stain it, and then count the annual growth lines. It is rather like counting the rings on a tree stump.

## Does Handling Affect the Bears?

Both we and the Inuit were anxious to learn if drugging and handling of polar bears caused any significant changes in their behavior, or other aspects of their biology. For example, hunters in the eastern Beaufort Sea asked if tagged bears could still hunt seals. Their concern was based on a local belief that if a bear was badly frightened by a hunter, it would be too nervous afterward to be able to sit still when waiting for a seal to surface at its breathing hole. The approaching seal would hear the bear moving and go elsewhere to breathe. As a result, the bear would

starve to death. When hunters saw thin bears, they wondered if they had escaped another hunter at some earlier time.

Would bears chased by helicopters be too afraid to be able to hunt properly? In other areas, hunters asked if bears would leave the area after being tagged and not come back. These were important questions and we addressed them in two ways. First, in our behavior studies at Radstock Bay (see the chapter on behavior) we compared the behavior of marked and unmarked bears of the same age and sex categories (i.e., adult males, adult females with cubs, and so on). We found no differences in the number of seals killed or the amount of time spent in various other activities such as sleeping or walking. Nor did marked animals leave the area. In fact the bears seemed remarkably unaffected. They went about their business after waking up in much the same manner as they did before. From the longer-term studies of marked animals, it soon became clear that they continued to come back to the same areas year after year despite having been caught there several times.

Malcolm Ramsay and I then reviewed data on almost 3,000 records in the Polar Bear Project's computer files on captured animals, going back for about twenty years. We checked aspects we thought might be important such as body weights, litter size, survival of cubs, and behavior. We found no significant differences. Recaptured mothers appeared to be a little lighter but their cubs were not. Recaptured mothers of one- and two-year-old cubs tended to have slightly smaller litters, although the difference was not statistically significant. Finally, we found that several pregnant female polar bears caught in earth dens in October moved about 25 kilometers (15 mi.) to new den sites before settling down again for the winter. In total, the effects on females with cubs were not large and there was no indication that they were detrimental. Nevertheless, we have flagged these facts for further attention as we continue to try to improve our techniques.

## What Do You Need to Know for Conservation?

Once you have learned how to catch polar bears and tag and measure them, the next task is to figure out how to apply these techniques to different kinds of problems. For example, in most areas of the Canadian Arctic, Inuit hunters sell polar bear hides each year as an important part of their income. Obviously, it is important not to harvest too many bears from the population. Thus we need to know how many bears there are and such things as how long they live, how often the females have cubs, how many cubs they have in a litter and how many in a lifetime, and how long bears of different age and sex classes survive through the years. With this information, one can approximate the safe sustainable annual harvest.

Most studies of population size use the "mark and recapture" approach to garner needed facts. This means that you capture as many animals as you can in the first year and mark them individually. In the following year, you go out and capture another sample. Some of the animals will be marked and some will not. Then, you use the ratio of marked animals to unmarked animals in your second sample to estimate the size of the population. For example, suppose you tagged 100 polar bears the first year. In the second year, you caught 100 more, of which 10 had been tagged the previous year. This would suggest that 10 percent of the total population was tagged because 10 percent (10 of 100) of the animals you caught were marked. Therefore, since you caught 100 in the first year, and that was 10 percent of the population, the total would be 1,000 bears.

Unfortunately, like most things in life, the theory is simpler than the practice. For the mark-recapture method to be successful for estimating the size of animal populations, you must make a few basic assumptions. These are: (1) every animal in the population must have an equal chance of being caught in year one or recaptured in year two; (2) animals do not enter or leave the study area; (3) tagged and untagged animals have an equal chance of being captured; (4) tags on any recaptured animal can be identified. These conditions are not easy to meet when sampling a polar bear population because the different age and sex classes of bears often segregate themselves into different areas or habitats. For example, one area may have more adult males while another may have more females with cubs. To compensate for this potential source of bias, you have to spread out the capturing effort over all kinds of habitat so that every bear has an equal chance of being captured or recaptured. In the Beaufort Sea in the spring, we found that the preferred habitat for most bears was the area of active leads at the interface of the landfast ice and the moving pack. Seals were more abundant there, and perhaps easier to catch, so lots of bears of all ages and sexes were there but especially adult males. At the same time, in the stable, drifted pressure ridges, far from the floe edge, there were many more females with newborn cubs and some subadults. To accommodate this difference, we had to ensure an adequate amount of searching in all types of ice habitats.

## When to Study Polar Bears

To estimate populations accurately, you need a representative sample of all the bears in an area. Yet scientists are often limited in what they can do by the amount of money they have to work with. If money is short, then you must concentrate on the most productive time and place. Population sampling during winter is pretty well out of the question. Twenty-four-hour darkness makes it impossible to find bears with helicopters. Extreme cold is hard on equipment and people alike, making the work difficult. In areas like the Beaufort and Chukchi seas, or the Canadian Arctic Islands, most studies are done between late March and mid-May. During that time period, the most representative sample of the polar bear population is likely to be present on the coastal landfast ice and adjacent floes. There are several reasons for this. Females with cubs of the year have come out of their dens so that data on productivity and maternity denning can be collected. In most areas, adult females with yearling and two-year-old cubs, and subadults, seem to prefer the area where the fast ice and floes meet as well. Some of the adult females will be looking for mates, and this pulls in the adult males that are often farther offshore. Coincidentally, long days and reasonably stable weather conditions prevail, making it possible to get more work done. Finally, tracking conditions are best at that time so that a larger number of bears can be captured than is possible by simply trying to spot them on the ice.

In Hudson Bay, the situation is quite different. The sea ice is very broken along the coast, making it quite difficult to track and capture bears. During the summer, however, the sea ice melts so that the bears are all forced ashore for up to several months. On shore when the snow is gone, white bears are highly visible, so they can be found easily. Finally, they tend to be abundant in areas such as the coastlines of Ontario and Manitoba and on islands like Southampton, Mansel, and Coats. The days are long and the temperatures reasonable. Here, it is possible to do some of the most cost-effective work in the world on polar bears.

# Distribution and Abundance

Polar bears are found throughout the ice-covered waters of the circumpolar Arctic (see Fig. 1). They prefer to remain on the sea ice all year round if possible because they need a platform from which to hunt. Polar bears catch very few seals in the open water. So when the edge of the ice retreats to the north during the summer, the bears will travel many miles to remain with the floes, where they have access to the seals. Should a bear let itself be stranded on land, it must remain there until the sea ice freezes again in the fall. When the young ice forms again in the autumn, the polar bears quickly move out onto it to resume hunting.

The most favored habitat of the polar bear is in the annual ice that lies adjacent to the continental arctic coastlines or island archipelagoes such as are found in northern Canada, Greenland, or the Soviet Union. A few years ago, we did an analysis of observations of polar bears and their tracks collected in surveys of about 75,000 kilometers of the Beaufort Sea to determine more specifically what kinds of sea ice habitat polar bears preferred. Although we identified seven general categories, the great majority of the bears we found (514 of 627) were near the edges of leads in the ice or in areas where the ice regularly cracked open because of wind and currents and then refroze. Bearded seals and young ringed

seals are more abundant in these areas. They are accessible to the bears when they surface to breathe in narrow cracks or at breathing holes in patches of thin ice that have just frozen over. In those areas, the holes have not yet become hidden by the drifted snow. Subadult ringed seals are less experienced than adults, and probably easier for bears to catch. Places where subadult seals are more abundant are especially attractive to polar bears.

The importance of these cracks was rather dramatically illustrated a few years ago in early March when one of the most experienced members of our polar bear crew, Dennis Andriashek, went off to the Beaufort Sea to put some radio collars on female polar bears. The weather had been calm and very cold for some time before he arrived and the ice was solidly frozen. There were no fresh cracks and the polar bears were widely dispersed at low density. After flying over 2,500 kilometers (1,550 mi.) he saw only six bears. Then the weather changed and strong winds blew for several days, opening up a single new lead about 150 kilometers (95 mi.) offshore. In the next six days, he sighted ninety polar bears in the vicinity of the lead. Even allowing for possible duplicate sightings, that is a remarkable difference. The importance of the lead was further underlined by the fact that almost all the tracks

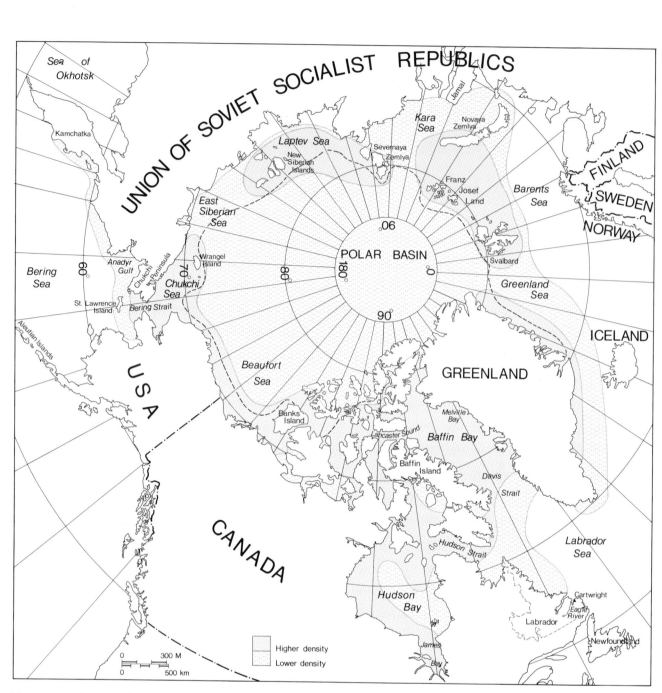

Fig. 1. Distribution of polar bears

recorded were in a corridor about 1 kilometer (0.5 mi.) wide on either side.

In stable ice areas where snowdrifts accumulate along the pressure ridges, female ringed seals maintain breathing holes in the ice below the drifts. In spring, they dig out lairs in which they give birth to their pups. About 100 of the 627 polar bears were found in this kind of habitat, including the great majority of female polar bears with cubs of the year. In most areas of the Arctic, this habitat tends to be along the coastlines, in large bays with islands in them, and in fiords

POLAR BEARS

such as those along the east coast of Baffin Island.

Polar bears generally avoid areas where the ice is very rough. Nor do they like smooth ice with little drifted snow on the surface to hide the breathing holes of seals. They also tend to avoid areas of multiyear ice such as that which characterizes much of the polar basin, probably because the density of seals is very low there. For these reasons, polar bears concentrate in the annual ice along the coastal areas to the south.

## The "Arctic Ring of Life"

Parallel to the coastline of the polar basin, and around the various archipelagoes, there is a system of leads and areas of open water surrounded by ice, called polynyas, which remain open through the winter. Wind, upwellings, and tidal currents keep the ice from forming. Recurring polynyas, ones that occur in the same place each year, are of the greatest biological importance because wintering or migrating marine mammals and birds can rely on them as breathing and feeding areas. Because of its biological importance, the Soviet biologist Savva Uspenski has given this zone a somewhat romantic sounding, though quite accurate term, the "Arctic ring of life" (see Fig. 2).

To date, there has been little research done on the biology of the ocean in shore lead and polynya areas, largely because they are often difficult to get to and dangerous to travel in. It is also very expensive to maintain a research project in such areas on a large enough scale to be worthwhile. However, for many years it has been well known to explorers and arctic biologists that there are substantial numbers of marine mammals and birds along the edge of the ice pack and in the shore leads and polynyas that parallel the arctic basin. They are there on a year-round basis, but especially during the summer. Some recent studies have suggested that the biological produc-

tivity of the ocean may be greater there as well. The wind and currents that physically maintain the polynyas also function to mix nutrients throughout the water column. The lack of ice on the surface allows more light into the water to stimulate photosynthesis and primary biological productivity. The coastal shore leads and polynyas also tend to lie over the shallower waters of the continental shelf, which, in general, are also more productive than the deeper oceanic waters offshore. Because the coastal areas are more productive, there is more marine life there of all kinds, including polar bears.

## Seasonal Changes in Distribution

Through the winter, areas of open water periodically appear and refreeze with changes in the weather; this influences the distribution of seals. Between summer and winter, the amount of ice-covered water can change radically and the bears must move with it. As a result, the life of a polar bear is one of constant adjustment to an ever-changing environment and a moving food source. This means that in some regions, such as the Bering or Greenland seas, the distribution of polar bears changes greatly through the year in response to seasonal changes in the distribution of sea ice. In other areas like Hudson and James bays, the ice melts completely in summer. The bears are forced ashore where they remain until freeze-up allows them to return and hunt seals again. These animals have less time to hunt, so they must locate the most productive areas. As summer nears again, they must deposit enough fat to see them through the ice- and seal-free summer months.

In intermediate areas, such as the archipelagoes of the Canadian Arctic or Svalbard and Franz Josef Land, bears may stay on the ice most of the time. Even so, in some years, they may spend up to a few months on land.

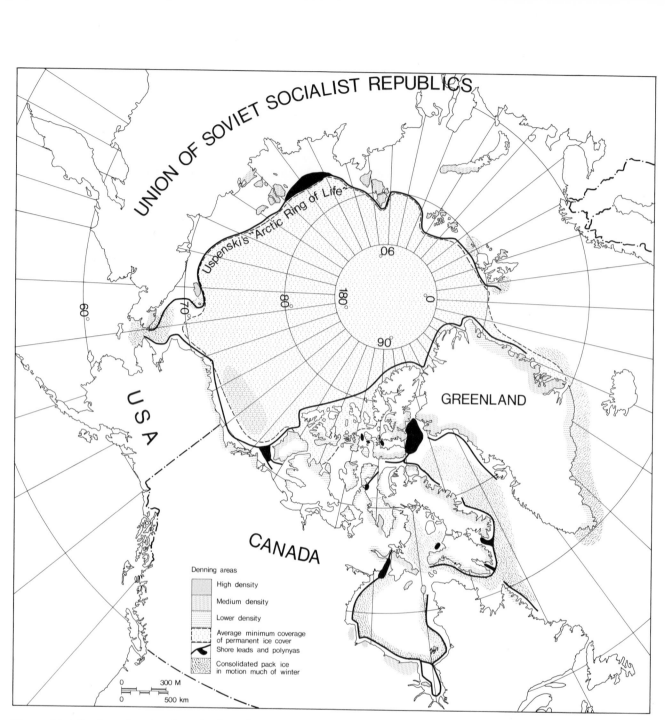

Fig. 2. Distribution of major shore leads, polynyas, and maternity denning habitat

Since the majority of polar bears have their maternity dens on land, they must adjust their pattern of seasonal movements in relation to changes in ice conditions so that suitable denning areas are accessible to them at the appropriate time.

## The Southern Limits

The southern boundaries of the circumpolar range of the polar bear are determined by the distribution of the pack ice during the winter. In the Bering Sea, for example, there may be little or no ice during the summer, so the polar bears move north as the edge of the pack ice recedes into the Chukchi Sea. During the winter, huge areas of sea ice freeze in the northern Bering Sea and then drift south several hundred kilometers, and therefore polar bears are annual visitors at St. Lawrence Island. They have been recorded as far south as St. Matthew Island and even in the Pribilof Islands to the southeast in exceptional years.

On the southwestern side of the Bering Sea, along the coast of Siberia, polar bears have occasionally been carried on the drifting ice floes as far as Kamchatka. On rare occasions, in particularly heavy ice years, bears have shown up on the coast of Hokkaido. According to Savva Uspenski, it is quite common for bears to be stranded on the south side of the Chukchi Peninsula when the ice melts in the northwestern Bering Sea during summer. Once on land, apparently, some of these bears walk cross-country, up to several hundred kilometers inland, as they travel from the Anadyr Gulf back to the Chukchi Sea to the north.

In the Labrador Sea, which bounds the northwestern Atlantic Ocean, polar bears live in the pack ice thoughout the year. They move north with the receding floe edge in the summer and south again in the winter when the seas refreeze. Along the coast of southern Labrador and off the north coast of Newfoundland, polar bears arrive with the pack ice each winter. In most years a few make it as far as the famous pupping grounds of the harp seal. The presence of a few hundred thousand harp seals and their whitecoat pups must about as close to polar bear heaven as it is possible to get. The bears sometimes kill large numbers of pups yet do not eat them all. In the years prior to the recent cessation of commercial sealing, one or two bears were killed each year by the seal hunters. Occasionally bears also reach the island of Newfoundland. In the heavy ice year of 1973, four bears

(two males and two females) came ashore along the northeast coast and were subsequently shot.

In the Barents and Greenland seas, which lie along the northeastern perimeter of the Atlantic Ocean, the distribution of polar bears is also governed by the seasonally changing distribution of sea ice. The Norwegian polar bear trappers in Svalbard knew that the number of bears present each year was largely determined by the presence of pack ice around the various islands. The bears arrived with the floes as they drifted south in the fall and left with them again in the late spring. In most years, relatively few bears remain on land in the islands of Svalbard because they prefer the ice floes where they can still hunt seals. In some years, however, a large field of ice remains attached to the land on the south side of Køngsøya in eastern Svalbard. When that happens, there are usually large numbers of bears around the island through the summer as well.

There is probably a similar pattern of movements of polar bears north and south with the ice around Novaya Zemlya in the eastern Barents Sea, but this is not well documented. Periodically a few polar bears arrive on the north coast of Iceland with the pack ice during winter, although this is not common.

The most southerly dwelling polar bears in the world live all year round in James Bay in Canada, at latitudes as far south as Calgary, Alberta, or London, England. There, and in Hudson Bay to the north, the ice melts completely in summer. There is no ice anywhere else to migrate to so the bears simply spend the summer along the coastline, or on various of the islands, waiting for winter to return.

## The Northern Limit

Polar bears or their tracks have been reported, albeit infrequently, by various explorers almost as far north as the pole and at other locations deep in the polar basin. This is uncommon, though, because the heavy multiyear floes that characterize the central Arctic Ocean are not much inhabited by polar bears.

The Norwegian polar bear biologist Thor Larsen and his colleagues conducted a series of aerial surveys in the Greenland and Barents seas during the early spring, and ship surveys in the summer. Their objective was to determine the northern limits of the distribution of polar bears in those areas. A very small proportion of the tracks they recorded from the air, and only 3 of 181 bears seen from the ship were above 82° north latitude.

In 1969, the Transpolar Expedition traveled from Alaska over the pole and south to Svalbard in the spring. They, too, reported to Larsen that they saw no bears between the pole and 82° north latitude, but found them in abundance farther south.

The northern limit of polar bears in the Greenland and Barents seas coincides with the boundary between the relatively shallow waters over the continental shelf and the edge of the deep Arctic Ocean. It is also the area where the biologically rich waters of the North Atlantic current mix with the cold water flowing south from the polar basin, increasing its overall biological productivity. North of the convergence, the Arctic Ocean is relatively sterile and this probably explains why there are few polar bears there.

The edge of the continental shelf off the north slope of Alaska and along the western side of the Canadian Arctic Islands is fairly close to shore. In this area, most of the polar bears are found within a few hundred kilometers of the coast at any time of year. North of the Soviet Union, between the Barents and Chukchi seas, the continental shelf extends up to 800 kilometers (500 mi.) offshore in places. How far offshore polar bears are

found on a regular basis in those areas is not known.

## Movements

Several years ago, it was thought that all the polar bears in the Arctic might be part of one circumpolar population nomadically roaming at will over the arctic wastes. This notion of the cosmopolitan arctic citizen randomly visiting any or all of the polar countries at will through its lifetime had a certain romantic attraction. It was in vogue for a time, but basically it ignores most of what we know about the distribution and movements of mammals in general. When you think about it in an evolutionary or survival context, it does not make a lot of sense for an animal just to wander off in any direction on the chance that food, mates, denning habitat, and other necessities will automatically be found when needed. Human explorers of the Arctic who pursued this naive assumption usually met with a variety of disasters, including starvation. However, the theory persisted and it was essential to test it. The circumpolar arctic countries needed to determine to what degree they might share populations, and thus be jointly responsible for their conservation.

In the last twenty years, several thousand polar bears have been tagged by scientists from all the arctic countries. By recapturing these animals in subsequent years or by return of ear tags from animals shot by Inuk and Indian hunters, we have learned a great deal about the movements of polar bears in different parts of the Arctic. In more recent years, some additional movement information has also been obtained by following bears with both conventional and satellite radio collars.

The first clear result from tagging individual polar bears was that when biologists went back to a particular area in the same season of subsequent years, they tended to find many of the same animals. This behavior, called seasonal fidelity, means an animal is faithful to the same area in the same season every year. It is, however, the exact opposite of what the theory of a single population roaming the entire Arctic would predict. For example, in the mid-1960s, Jack Lentfer, formerly of the U.S. Fish and Wildlife Service, began his pioneering polar bear tagging studies on the sea ice north of Point Barrow, Alaska. Within a year or two, he began recapturing the same animals and Inupiat hunters in the same general area returned tags from bears they had killed.

Similarly, in the deep fiords of the east Greenland coast in the mid-1970s, the Danish biologist Christian Vibe caught and recaptured several polar bears in exactly the same bays where he first tagged them during the spring a year or two earlier.

In the late 1960s, Chuck Jonkel, formerly with the Canadian Wildlife Service and now at the University of Montana, found the same high degree of seasonal fidelity being shown by polar bears on the western coast of Hudson Bay in the autumn, as others were finding in the spring.

Several studies, completed all over the Arctic, have shown that an individual bear may be tagged and recaptured a year or more later at points that may range from a few kilometers to a few hundred kilometers (up to a hundred or more miles) apart. Studies of polar bears with satellite radios have shown that even females with cubs a few months of age are capable of traveling more than 30 kilometers a day for several days in a row. Although distances of a few hundred kilometers sound large, they represent normal movements for a polar bear within a general area during the course of a year. For example, even though bears captured in the Beaufort Sea move around quite a bit within the region during the year, they were not caught in the High Arctic Is-

lands of the Canadian Arctic Archipelago, or Siberia, or vice versa.

Even more remarkable was the separation of subpopulations in places where it seemed there was no natural barrier. For example, the bears of northwestern Ontario and James Bay appear to form a separate subpopulation from those in western Hudson Bay. Similarly, in the Beaufort Sea, there are two subpopulations: one that lives along the mainland coast approximately between Point Barrow and Baillie Islands and another that lives along the west coast of Banks Island and in Amundsen Gulf.

In April and May, the sea ice in the Beaufort Sea is a continuous sheet of ice for hundreds of kilometers, broken only by occasional leads in the ice. In most places, once the bears are more than a few kilometers offshore, there are no landmarks to navigate by, or at least none that are readily apparent to humans. In the summer, when the ice in the southern Beaufort melts, the bears must move several hundred kilometers to the north for a few months in order to remain on the pack ice. In the late fall, they return to the better seal hunting areas over the shallower water closer to the coast after the ice refreezes. In addition, the sea ice moves constantly at speeds up to several kilometers per day, depending on the season. The pack ice circulates in a gigantic clockwise pattern called the Beaufort Gyre. It drifts south from the polar basin along the west coast of Banks Island, then west along the Canadian and Alaskan coast to just past Point Barrow before it heads back north toward the pole again to begin another circuit. Bears must continuously compensate for the Beaufort Gyre in order to stay in the same place and, somehow, they do it. I have often recaptured polar bears out on the ice of the Beaufort Sea, far from sight of the nearest land, within a few kilometers of where they were first caught two, three, or even more than ten years previously.

In his landmark study of the movements of female polar bears in the Beaufort Sea, the American biologist Steve Amstrup radio tracked a few dozen individual female polar bears for up to four years. He found that some bears had huge home ranges that went back and forth along the whole southern coast of the Beaufort Sea while others stayed pretty much in the same place. His results show more clearly than any other study that the bears know exactly where they want to go; they are not just wandering aimlessly.

On the western coast of Hudson Bay we have been capturing and recapturing polar bears during the ice-free period from late July to early November for about twenty years. Some of the bears have been caught at the same general location in so many different years that they now seem like old friends. Even between the adjacent Provinces of Manitoba and Ontario, there has been remarkably little exchange of tagged polar bears over several years.

Malcolm Ramsay and Dennis Andriashek examined the direction taken by the tracks of seventy-four adult females with cubs as they left the maternity denning area south of Churchill in February and March. They found that the family groups took fairly straight routes to the sea on courses that were parallel to each other. The intriguing aspect was that the mean course taken by the bears was 39° true (close to northeast) while the most direct route would have been to go straight east.

## Home Ranges and Subpopulations

From these observations, you have to conclude that polar bears are good navigators, although we do not yet understand how they do it. Furthermore, they use that ability to remain within their individual home ranges and maintain contact among their subpopulations.

The term home range means the area within which an animal confines its movements. The size of a home range can vary greatly between species or even among individuals within a species. One factor influencing the size of the home range of polar bears is the annual pattern of freeze-up and breakup of the sea ice. A bear's home range may be small, if it has constant access to ice, leads, and seals. Because it does not need to move far, it stays in one place. On the other hand, in areas such as the Barents, Greenland, Chukchi, or Bering seas, the bears may have to move many hundreds of kilometers each year just to remain on the ice where they can still hunt seals.

Within their home ranges, bears must respond to seasonal fluctuations in the population and distribution of seals. For example, where there are many leads, seals move around quite a lot under water. Consequently, polar bears have nothing to gain by defending specific geographic areas, as do their brown and black bear cousins. On land, berry patches and ground squirrel colonies do not move around, and a dominant animal can benefit by staking out a claim and defending it. Seals, in contrast, may be in different areas between years.

Each individual polar bear within a subpopulation, except for cubs with their mothers, determines its own home range. A bear probably learns the seasonal pattern of movements as a cub during the two years or more it remains with its mother. As it gains in knowledge and experience, the bear builds on its mother's teachings to develop its own individual patterns of movement within its home range.

Since the bears' movements and home ranges are essentially independent of one another, subpopulations are concentrations of polar bears with independent but overlapping home ranges, in other words, a continuum of home ranges. The distribution of these home ranges, however, is neither random nor spread evenly throughout the Arctic. Ecological conditions and the distribution of resources are too variable to permit this. Thus, in places where ice conditions or low food abun-

When getting out of ice too thin to walk on, the bear spreads his huge feet far apart and sometimes crawls on his elbows and knees to spread his weight and keep from breaking through. As he gets to more solid ice, he begins to stand up and then walks normally again.

70

Arctic foxes follow polar bears on the sea ice to scavenge the remains of seal kills.

This old starving male still has enough energy to chase a fox away from his kill.

A subadult bear sits down to enjoy a piece of kelp.

Bears sometimes like a mouthful of grass. Most grass is eaten earlier in the summer when it is still green and growing.

TABLE I.  **Estimates of the Size of Subpopulations of Polar Bears**

| Area | Basis of Estimate | Source | Estimate |
|---|---|---|---|
| Barents and Greenland seas | shipboard surveys | Larsen | 5,000 |
| East Greenland | mark-recapture and aerial surveys | Vibe | 200 |
| Baffin Bay and Thule district | subjective, based on harvest | Vibe | 300 |
| Canadian Arctic | | | |
| Zones A1 and A2 (southwestern Hudson Bay and James Bay) | mark-recapture and aerial surveys | Kolenosky Jonkel | 1,500 |
| Zone A3 (western Hudson Bay) | mark-recapture | Stirling | 1,500 |
| Zone B (Labrador coast) | mark-recapture | Stirling | 75 |
| Zone D (south) | mark-recapture | Stirling | 700 |
| Zone E (Central Arctic) | mark-recapture | Schweinsburg | 1,100 |
| Zone F (High Arctic) | mark-recapture | Schweinsburg | 2,000 |
| Zone H (eastern Beaufort Sea, Amundsen Gulf) | mark-recapture | Stirling | 1,200 |
| Alaska North Slope and western Zone H | mark-recapture and other indices | Amstrup | 2,000 |
| Soviet Union | aerial survey | Uspenski | 3,600 |

dance preclude seals, there will be fewer polar bear home ranges. Conversely, in areas where biological productivity is high and seals are abundant, there will be many overlapping polar bear home ranges. Together, they form what we would call, for conservation purposes at least, a subpopulation.

## How Many Bears Are There?

This must be the most frequently asked question about polar bears. It recurs because people think that polar bears are an endangered species. They are not. Even so, we do not know the size of the total world population of bears. And even if we did know exactly, it would not be particularly useful in conserving or managing the discrete subpopulations. To do that, we need to know the size of separate subpopulations and their ranges.

Research on the population ecology of polar bears is both time-consuming and very expensive. Although we have estimates of the sizes of several subpopulations, we are far from having even an initial assessment of all areas. Even in

some areas where research has been conducted, the estimates of population size can only be called educated guesses.

Worldwide concern for polar bears has led scientists to concentrate on population estimates, even though lack of funds has limited their ability to do the work. However, at the Eighth Working Meeting of the IUCN Polar Bear Specialists Group, held in Oslo, Norway, in 1981, there was general agreement that the world population is certainly greater than 20,000 and could be as high as 40,000. Personally, I am inclined toward the upper end of that range.

At the same meeting in Oslo, there was also a general review of the sizes of some subpopulations in the Arctic. Shown in Table 1, they are not absolute numbers, just the estimates of those involved. The total of 19,175 represents estimates just for those subpopulations for which surveys were performed. It is impossible to compare estimates made by such variable methods as mark-recapture, aerial counts, and shipboard surveys. Besides not being easy to see on ice or snow, polar bears are widely spread out over huge areas at such low densities so that it is difficult to get large enough sample sizes, no matter what tech-

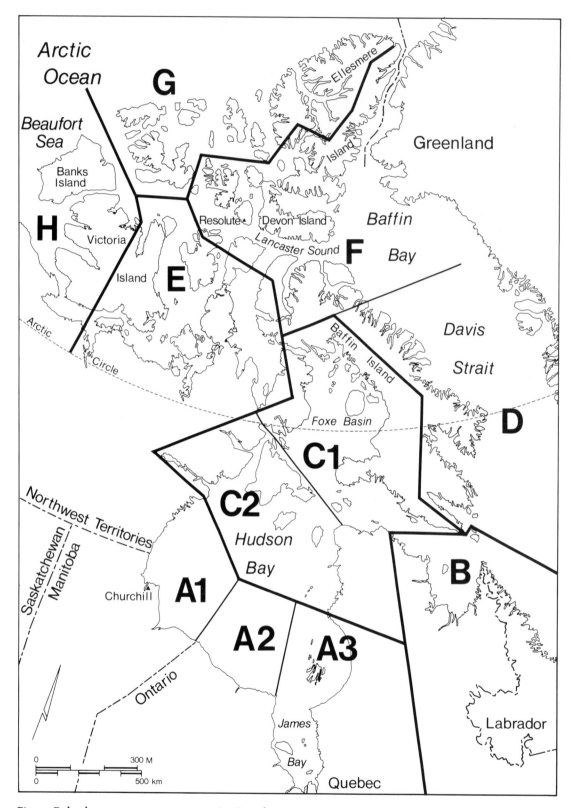

Fig. 3. Polar bear management zones in Canada

nique is applied. There are many areas where polar bears are present but not represented in this total. In Canada alone, this includes Foxe Basin (Zone C) where more than 130 polar bears are harvested each year, the east coast of Baffin Island (Zone D north), and the whole of the northwestern High Arctic (Zone G). Figure 3 shows the location of the management zones in Canada. These are based on existing knowledge of the distribution of subpopulations; the boundaries are revised periodically as new results become available. Even within some of the management zones studied, the intensity of the sampling varied geographically, so that there are probably more bears than our mark and recapture estimates suggest. In some cases in which little research has been done, the estimates are simply the educated guesses of the scientists familiar with the areas.

Similarly, we have no reliable estimate for the size of the polar bear population in the Chukchi Sea, although we do know that for many years bears were taken there in large numbers by Alaskan hunters. Some scientists have approached the problem by making estimates based on mark and recapture data. Others have used computer simulations to estimate how large the population would have to be to sustain a known harvest over the years. Different estimates range from 2,500 to 7,000. Again, these are just educated guesses, but it is clear that there are a lot of polar bears there.

The largest gap in our knowledge has to do with polar bear populations in the Soviet Union. Most Soviet research has been aimed at ecology and behavior, especially that of denning females.

The Soviets have done no mark and recapture studies similar to those done in Canada or Alaska. The Soviets have carried out extensive aerial surveys; our experience with this approach has been disappointing, however. The animals are just too difficult to see from the air. For example, one sunny day, an experienced pilot with very good eyesight and I were flying right along the floe edge about 60 meters (220 ft.) above the ice. We were looking for bears in a fairly narrow swath right in front of us. Something dark caught my eye as it passed beneath us and I thought it was the remains of a seal killed by a bear. We turned to collect specimens and saw two large male polar bears standing feeding on a ringed seal. I was amazed that two experienced people looking for bears at such a low altitude did not see them. We have had a number of similar experiences to convince us that trying to count polar bears on the ice from aircraft is a waste of time.

On the basis of their aerial surveys, Uspenski and Shilnikov estimated there were 1,800 to 3,600 bears in the Soviet Union. (In Table 1, I deliberately used the 3,600 figure because their numbers seem unrealistically low.) Given that about half of the Arctic lies within the Soviet Union, and that an incomplete assessment of population size for the rest of the Arctic probably exceeds 20,000 (15,575 from Table 1 plus the unsurveyed areas listed above), it seems likely that the Soviet figures greatly underestimate the number of polar bears in that country. When all this is considered, it seems quite possible that the world population of polar bears could be approaching 40,000.

# Reproduction

Polar bears behave differently from black and brown bears because of differences in their habitat and variations in other ecological factors. For example, the land bears defend territories because the best areas for feeding are always in the same place each year. The territories of adult males are usually larger than those of females. Each male territory may overlap part or all of several adult female territories. Because of this, a male often has the opportunity to mate with more than one female. As a result, land bears may live their entire lives within a fairly small area, since there is little reason to roam elsewhere.

A territorial male can influence access to females by competitors. This results in intense competition between males for mating rights. Bigger individuals win more fights and consequently do more mating, and thus there is a strong evolutionary impulse for larger body size. This leads to "sexual dimorphism," which means a pronounced size difference between males and females. Some of the most extreme examples of males becoming larger than females, as a result of competition for reproductive advantage, are the terrestrial-breeding seals such as the elephant seal or the northern fur seal.

The sea ice habitat of the polar bear differs from that of the rest of the bears in one major way: it is highly variable between seasons and years. For example, because of a different pattern of freeze-up or breakup, there might not be any suitable ice for hunting seals in an area where they were abundant just the year before. Obviously this is an extreme (though not unusual) example. Still, it is quite common for the location of the best seal habitat to shift a hundred or more kilometers between years even though it remains in the same general area. Consequently, polar bears are not territorial. Males and females meet and mate on the sea ice wherever the seal hunting is best.

Although the overall ratio of male polar bears to females in the population is approximately even, only about one-third of the females are available for reproduction in a particular year. This is because in most areas of the Arctic, females keep their cubs until they are at least two and a half years of age, so they only breed once every three years. This means, on average, that there are three available adult males for each breeding female. This ratio produces such intense competition between males for mating rights that polar bears have become one of the most sexually dimorphic of all mammals. Dominant males are two to three times the size of females. In other species in which sexual dimorphism is prominent, there are usually far fewer males than females. The reverse is true of polar bears.

Another source of evidence for the intense competition that takes place between males is recorded in their teeth. Old adult males often have badly broken canines. If this was just a result of old age and wear from a lifetime of hunting seals, one would expect to see similar breakage in females, but it is not present. The most likely interpretation is that the males break their canine teeth during intense interactions with each other as they compete for females during the breeding season. Males also show battle scars on their heads, necks, and shoulders. Where the white hair has been torn off, the black skin shows through.

## Breeding Behavior

Females do not usually mate when they are accompanied by cubs of the year or yearlings. It seems likely that the hormones associated with lactation (the secretion of milk) also inhibit ovulation. By the time the cubs are two and a half years old, lactation has ceased. The female either loses her cubs herself or they are frightened away by an adult male when he begins to follow the female for mating.

Mating takes place on the sea ice in April and May. With the exception of family groups, polar bears normally live a solitary existence. Just finding a mate at the right time could pose a significant problem. The concentration of adult males and females in the best seal hunting habitat sets the scene for encountering the opposite sex. At this time of year, adult males travel extensively in search of reproductive females. I have followed the tracks of individual adult males as they plodded relentlessly in a more or less straight line across the frozen pack ice, or along the edge of a lead, for 100 or more kilometers (60 mi.) in search of a breeding female. I don't know how a male discerns the track of a breeding female from the dozens of other sets of footprints he might

cross in a day, but recognition is instant. He immediately begins to follow her tracks, regardless of what direction she is going. It is possible that there is some chemical signature being given off in her urine or maybe even through specialized cells in the soles of her feet, but at present this is unknown. Regardless of how she does it, the male polar bear is never in doubt. He follows her tracks wherever they go until he catches up with her. At first, she may be accompanied by more than one male. Considerable fighting follows before the dominant male chases his rivals off. We have seen as many as five or six additional males in the general vicinity of a single male-female pair. That is probably the sort of intense competition that produces the broken teeth and scars referred to earlier.

Once a male has secured a female, they appear to stay together for a week or more. He will move her away from the prime feeding habitat into secluded bays, high up on mountainsides, or out on the pack ice away from the prime feeding areas. This lowers the chance of encountering other males. It is easy to recognize the tracks of a male-female pair because they go back and forth all over the area and join where they have stood and mated.

The fact that they stay together for so long is another clue to their reproductive biology. Some mammals, such as humans or seals, are "spontaneous ovulators," meaning that the females automatically release an egg for fertilization at the appropriate biological time because there is a high degree of certainty that there will be a male available to fertilize it.

The situation is quite different for a female polar bear. Because the adults are usually widely distributed as solitary animals at fairly low densities, there is no guarantee there would be a suitable male nearby on the day she ovulated if she did so spontaneously. So, although it has not been demonstrated experimentally, what probably happens is that the female comes into a state of physiological readiness for mating but does not actually release the egg for fertilization until the stimulation of mating causes her to do so. This process is called induced ovulation, and it probably occurs in terrestrial bears as well as polar bears.

In polar bears, several days of interactions and mating may be required to stimulate ovulation. I have watched male-female pairs traveling together and have seen how the male constantly tests the female to see if she is willing to mate. At different times, she may ignore him, run away a short distance, be aggressive and chase him away a short distance, or tolerate close contact. They may alternate periods of intense interaction with occasional lapses into hunting for seals or sleeping near each other in pits in the snow. From looking at the tracks of many pairs, it is obvious that once they begin to mate, they do so many times. Multiple matings over several days are probably required to stimulate ovulation and ensure fertilization.

Some observations made in the Moscow Zoo in the 1950s are particularly interesting because they are so consistent with what we have seen in the field. A single male bear was being held with a small group of adult females. The male suddenly began to follow one of the females for several days in late February and early March, although she continued to ignore him. Then, on the third of March, they began to mate and did so periodically for nine days before she again rebuffed him. He continued to follow her for another two days without mating before losing interest. On March 15, he began to follow another female and mated with her for four days beginning on March 30. The following year, he mated with a third female for twelve days, from the fourteenth to the twenty-sixth of March. The durations of these matings in captivity were not recorded, but three that were observed in the wild at Wrangel Island lasted 3, 7, and 10 minutes.

The length of time the male stayed with the females in the zoo is similar to the few observations we have from wild polar bears, plus what we can interpret from their tracks. Together, these data support the idea that polar bears are induced ovulators and require considerable stimulation before the egg is released for fertilization. There is another benefit to this reproductive strategy besides simply ensuring that the egg is not wasted. Since the pair remains together, interacting and mating, for several days before the female ovulates, this allows time for intense competition to take place between the adult males before her eventual consort is determined. This increases the probability that her cubs will be sired by a large, dominant, successful individual. This way, they will receive the best possible genetic inheritance, which will give them the best chance of surviving to pass along their superior genes to the next generation. It also means the most dominant males may succeed in mating with several females in each breeding season, which further emphasizes the advantages of being bigger, stronger, and more aggressive than one's competitors.

## Getting Ready for Denning

After the female mates in the spring, she has only a few short months in which to store away the large deposits of fat she will need to live on and support her new cubs after she enters her maternity den in the coming fall. This is no mean feat, since adult females weigh only about 150 to 175 kilograms (330 to 385 lb.) when they wean their cubs. They need to gain at least 200 kilograms (440 lb.) of fat to carry off a successful pregnancy. One of the most dramatic examples of how fast an adult female can change weight was provided by an individual that was caught with two yearling cubs in the Churchill dump in early December. She was a bag of bones and weighed only 97 kilograms (213 lb.). By the next spring she either weaned or lost her cubs, and mated. When we recaptured her the following August, eight months later, she weighed over 450 kilograms (992 lb.). Her fourfold increase in weight was largely due to fat deposition. The following summer, she was recaptured with triplets.

The timing of both mating and the weaning of cubs has probably evolved to occur in the spring. That is when the seals they eat are most abundant and vulnerable to predation. In late March and early April, ringed seals give birth to their pups in the tens of thousands. The female seals dig birth lairs in snowdrifts over breathing holes in the ice. At birth, pups weigh about 4 to 5 kilograms (10 lb.). Six weeks later, when they are weaned, they weigh 25 to 30 kilograms (50 to 60 lb.). At weaning, anywhere from about 40 to 75 percent of their total body weight can be fat. At this young age, they have not learned to avoid predators, so polar bears feed on them intensively through the spring and early summer. It is this superabundance of fat, naive seal pups that enables the pregnant females to accumulate fat so quickly. Later in the summer, after breakup, the young seals disperse in the open water where predators cannot catch them. At that time, polar bears weigh more than at any other time of year.

Female polar bears sometimes eat a bit of vegetation after coming onto land in the fall. They may do so again just after breaking out of the den in the spring, but it seems unlikely vegetation contributes much to their overall energy requirements. Mainly, they live on their stored fat reserves.

How long a pregnant female must go without feeding can vary considerably between areas. In the Beaufort Sea for example, a female can remain with sea ice somewhere in her home range throughout the year. She can continue to catch seals until it is time for her to begin to dig a maternity den in about mid- to late October. In parts of the eastern Canadian Arctic, East Green-

land, and possibly the New Siberian Islands, most bears must spend at least some time on land during the open water period in the summer. However, because freeze-up in these areas starts by late September to early October, pregnant females have time to catch a few additional seals before going into their maternity dens. At Svalbard or Wrangel Island in the Soviet Union (not to be confused with Wrangell Island in Alaska), pregnant females continue to hunt in the drifting pack ice and do not come ashore until late August or September when the floes drift south toward the islands. Females in these areas spend about six months on land, including the time spent in maternity dens. In Hudson Bay, however, the sea ice melts by the end of July, forcing all the bears ashore. Since freeze-up does not occur until November, there is no opportunity for pregnant females to feed on seals again before they enter their maternity dens. Many of these females must live for eight months on their fat reserves, give birth to cubs (usually two), and nurse them up to a weight of 10 to 15 kilograms (22 to 33 lb.) before leaving their dens for the sea ice again.

## Finding and Making Maternity Dens

No one has ever completely documented the behavior of pregnant females from the time they first come ashore through the period of selecting a den and remaining there through the winter. The Soviet scientist O. B. Lutziuk recorded observations of the behavior of possibly pregnant females in the early fall at Wrangel Island, in eastern Siberia. Substantial numbers of polar bears den there every year. They are relatively easy to observe when they first come onto land from the sea ice and walk about, possibly seeking suitable den sites. In most years, polar bears move off the pack ice onto the land in late August. If there are snowdrifts present from the previous winter, the bears dig dens in them right away. The females seemed to prefer the dense hard snow of the older drifts to the softer snow of freshly formed snowbanks. Lutziuk also suggested that pregnant females will reenter old maternity dens if they are present and remain in them through the winter. Although his observations only continued until about mid-September, some females that entered old dens in late August remained there through at least the first few snowstorms of the early fall.

The females he thought were settling into maternity dens were not displaced from them despite repeated visits, at least up to mid-September. The bears did not exhibit any signs of aggressive behavior toward the biologists on these visits, even when their dens were probed with wooden poles to ensure they were still occupied. In contrast, bears in what he thought were temporary pits were easily frightened when approached.

Ray Schweinsburg, formerly the polar bear biologist for the Northwest Territories Wildlife Service, found quite a different situation on northeastern Baffin Island in the Canadian High Arctic. During August of 1975 and 1976, he found sixty-three dens at altitudes ranging from 33 to 660 meters (108 to 2,165 ft.). In 1975, none of the snowdrifts were composed of fresh snow, whereas in 1976 many were. In 1976, all the dens found were in new snowdrifts. However, many of the dens found by Schweinsburg were occupied by adult males, and females with cubs of various ages. Clearly, they were not going to be denning for the winter. He particularly noted how fat the bears in these summer dens were and speculated that the bears were conserving energy during the open water period when seal hunting was not practical. Although pregnant females might remain and use the dens for having their cubs in as well, it has not been documented.

Lutziuk could not be sure the females he saw digging dens in old snowdrifts, or reoccupying old maternity dens, remained in them throughout the winter, because he was not able to monitor them through the whole autumn period. However, some of his observations are reminiscent of things we have seen pregnant females do in southwestern Hudson Bay in August and September. Of course there is no snow on the ground there but there is permafrost along some of the streambeds and edges of lakes under copses of black spruce trees. There are many old dens dug in the earth against the permafrost, and pregnant females reuse many of them for maternity dens (see Fig. 4). We have also noted that once into these dens, even in the early fall, the females do not want to come out and are fairly unaggressive when approached to close range. Recalling that pregnant females at Wrangel Island and southwestern Hudson Bay must fast for longer periods of time than polar bears in other areas, it seems likely that the function of this early fall denning behavior is to conserve energy. Just as we know that pregnant females in Hudson Bay remain in their earth dens until snowdrifts form over them, it is plausible that the females observed by Lutziuk remained where he saw them throughout the winter. However, it remains unconfirmed.

Pregnant female polar bears that spend the summer on the pack ice hunting through the summer and fall before denning behave quite differently. They have no need to look for an interim resting place. When they come ashore, in late October, they do so to look for a maternity denning site. Their tracks are unmistakable as they walk from snowbank to snowbank, testing them for consistency or depth or whatever else is critical to pregnant female polar bears. From the number of test pits and holes they dig, it is obvious that they know exactly what they are looking for, and they may travel many kilometers in the search. Provided there are suitable drifts around, a pregnant female probably selects a site within a few days.

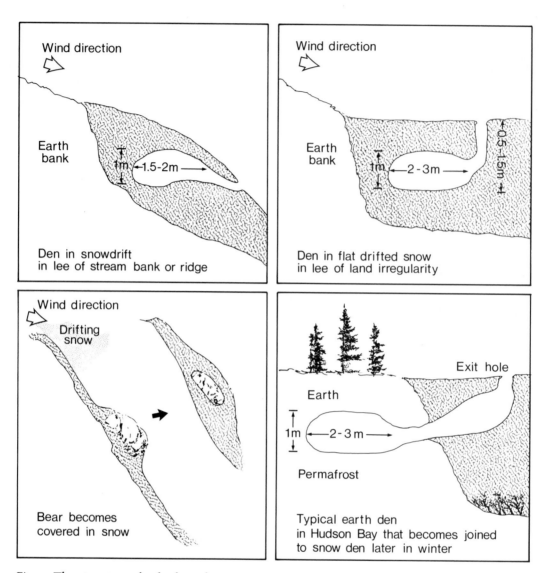

Fig. 4. The structure of polar bear dens

In his classic work on the denning habits of polar bears in the Canadian Arctic, Richard Harington, the original polar bear biologist for the Canadian Wildlife Service, gathered data on over 100 maternity dens to find out what influenced the selection of sites by pregnant females. In the Canadian Arctic Islands, 61 percent of 113 dens were within 8 kilometers (5 mi.) of the coast and 81 percent lay within 16 kilometers (10 mi.). Similar distances from the coast were reported from Svalbard and Wrangel Island. In contrast, along the southwestern coast of Hudson Bay (in Manitoba and Ontario) maternity dens are an average of 50 to 60 kilometers (30 to 35 mi.) inland from the coast. Some are as far as 100 kilometers (60 mi.) or more. The different terrain may explain the variation. Much of the coastline and adjacent inland areas of southwestern Hudson Bay is flat and boggy. There are few banks high enough to contain earth dens or deflect blowing snow to form suitable drifts. Consequently, pregnant females probably have to go farther inland to find suitable denning habitat.

Harington found a tendency for maternity dens in the Canadian Arctic to face in a southerly direction where the northerly prevailing winds deposit the best drifts. There, too, the benefits of solar radiation are the greatest. In general, this is also true in southwestern Hudson Bay and on Svalbard.

On Wrangel Island, the distribution of snowdrifts seems to be quite variable, depending on which direction the winds came from in a particular fall. Consequently, the direction the dens faced varied considerably between years. A tendency for some dens to face north was thought to be influenced by the regular use of old snowdrifts from the previous winter, which, of course, last longest on the north and east facing slopes.

Observations concerning den sites in different areas may seem contradictory, but the annual regime of snowfall could account for it. In places like Southampton Island, the east coast of Baffin Island, or the southwestern coast of Hudson Bay, there is usually quite a bit of snow. So, although there may be some variability in when it falls, there will be lots of it and drifts will form that face in all directions. Consequently, pregnant bears can choose a drift that faces whatever direction they prefer. The information available suggests that more females dig dens on southerly facing slopes.

The High Arctic, in contrast, is a polar desert, which means that the total precipitation received in a year does not exceed what might be expected in a normal desert in more southerly climes. Because there is so much less snow, the distribution of the drifts is more strongly influenced by the direction of the prevailing winds. As a result, there can be considerable variation between years in the number of drifts present, and in their direction of orientation. On Banks Island, in the western Canadian Arctic, I have found dens facing all directions. On one particular hillside, I found dens in perfect-looking drifts in two springs while in six others there was so little snow you could see bare ground. In places where the distribution of suitable habitat is that variable, the polar bears probably do not have as much opportunity to be selective as they would if snowdrifts were more abundant each year.

## Maternity Dens

Female bears everywhere construct dens that are quite similar. Most common is a single chamber, slightly elevated from the entrance tunnel so that the warmer air remains with the female and her cubs (see Fig. 4). Fourteen single-room dens measured by Richard Harington averaged about 2 meters by 1.5 meters (6.6 × 5 ft.) and a meter high (3 ft.). The entrance tunnels are usually about 2 meters long but very narrow, about 0.65 meters (2 ft.) in diameter. I have squeezed through some of these entrance tunnels myself and I am amazed that there is room for the female bears. Some females become more architecturally creative and dig two- and even three-room complexes. The cubs sometimes also dig small alcoves off to the side. In areas where the snow continues to accumulate, it may be necessary to scrape snow from the roof of the den to keep it thin enough for oxygen to continue to pass through. Some, but not all, authors report the maintenance of a small air hole for ventilation.

Harington poked a hole through the roofs of two occupied dens and lowered a thermometer in on a string. In one, the temperature was -9.9° C (14.3° F), about 21° C (37° F) warmer than the outside air. In the other den, the inside temperature was −17.8° C (0° F), which was 7.8° C (18° F) warmer than it was outside. The dens were being kept warmer by the body heat given off by the bears, and the insulating properties of the snow. Sometimes the roof of the den may have a layer of ice crystals indicating that on occasion the inside temperature can rise considerably.

## Growth of the Cubs

Female polar bears mate in late April or May, but the fertilized egg remains in a state of physiological suspension until about August or September. It then implants in the wall of the uterus and begins to grow. This process is called delayed implantation.

The pregnant female digs and enters her maternity den by about late October or early November, and the young are born somewhere between late November and early January. At birth, the cubs are quite small and undeveloped, especially in relation to the large body size of their mothers. The same is true of all the bears, including the panda. Newly born polar bear cubs weigh less than a kilogram (1 to 1.5 lb.). They have such fine fur that its very existence has sometimes

been overlooked, resulting in inaccurate reports of their being hairless.

A male and female cub born in the Topeka Zoo in 1971 weighed 706 and 570 grams (a little over 1 lb.) respectively. Uspenski reported that three one-day-old male and female cubs in the Leningrad Zoo weighed 725–840 grams and 650–730 grams respectively. The mean lengths of the male and female cubs were 30.2 and 28.9 cm (about 1 ft.). Even from these few observations it appears that the male cubs are larger than the females right from birth. The mother cuddles her tiny cubs close to her body to keep them warm in the shelter of the den. By the time the cubs leave their dens, the males already weigh about 10 percent more than the females.

The cubs' eyes are closed at birth and do not open until sometime later. The cubs in the Topeka Zoo opened their eyes at 26 days of age (compared to 30 to 31 days in Leningrad), and their canine teeth first became apparent at about 50 days. Uspenski also reported that the sense of smell was first apparent at about 50 days of age. Unsteady walking began at about 2 months of age in both zoos.

## Getting Ready to Leave the Den

The female nurses the cubs on her rich milk until sometime between late February and the middle of April, depending on the latitude. Mother bears break out of their dens later in the far north than they do in the south where it gets warmer earlier. By the time she breaks out of her den, the cubs weigh about 10 to 15 kilograms (25 to 30 lb.). They are ready to go to the sea ice with her to hunt seals. When the female first opens the entrance of her den to the outside world, the air temperature can be −25 to −40° C (−10 to −40° F).

The most extensive observations of the behavior of family groups at the time they break out of their maternity dens were made by the

Norwegian biologists Rasmus Hansson and Jorn Thomassen. They spent two springs observing polar bears at Køngsøya in Svalbard. This is a particularly interesting location because, like Wrangel Island in Siberia, there may be tens of dens in a small area. Several dens can actually be within sight of each other. This contrasts with almost everywhere else in the Arctic, where polar bear dens are widely distributed at low densities.

In general, the females did not break out unless the weather was favorable, although there were exceptions. After the entrance was first opened, it was still an average of two days before the cubs began to venture outside. Overall, during the period prior to departure, the cubs still spent 85 percent of their time inside the dens. The family remained in the general vicinity of its den most of the time but occasionally went on short exploratory walks during which the female did some digging and grazing. I have often seen tracks and signs of this activity in places as separate as Banks Island and the west coast of Hudson Bay. When eating lichens and grasses, the females leave large stools characteristic of this diet. At Køngsøya, Hansen and Thomassen noted that defecation and urination took place all over the area but there was no indication of a scent-marking or territorial function. Three times they saw females using their paws to scrape snow over their feces like a dog. On a number of occasions, they observed the females eating the feces of the cubs, which has also been seen at Wrangel Island by Uspenski and Kistchinski. Cubs have not been observed eating feces.

When outside, the cubs played a great deal, chasing and rolling over each other, fighting, and sliding. Although the female did not participate in this behavior, she always remained close by her cubs. Polar bear females are attentive mothers and they groom and touch their cubs with their muzzles often. Nursing took place periodically with the the female either sitting up or lying on her side. Often, this was done in a temporary pit dug in the snow (p. 42). Curiously, some families moved and temporarily occupied the abandoned dens of other families.

At Køngsøya, Hansen and Thomassen found that females could be divided into two groups according to how long they remained at their dens after first breaking out. Most remained for about twelve days, although there was also a group that seemed to stay for longer than fifteen days. The period spent around the den before departure is probably very important. It gives the cubs a chance to acclimate to the cold they must now endure after the shelter and relative warmth of their previous home. By March 10, most new families in the southwestern Hudson Bay population are gone from the denning area. In the Canadian High Arctic, 2,000 kilometers (1,300 mi.) to the north, females do not leave their dens until April, or even early May.

Finally, the female decides it is time and the family begins its trek to the sea ice. The periods of travel are punctuated with rests and bouts of nursing in resting pits scraped out of the snow. The cubs remain with their mothers for one and a half to two and a half years depending on the area, and sometimes even longer. Thus, females can only have cubs about once every two or three years or more.

## Age of Reproductive Maturity

Theoretically, all long-lived mammals born in a given year should reach sexual maturity at about the same time. Sexual maturity is the age at which a mammal is capable of mating for the first time. In practice, a good proportion of any cohort (i.e., all young born in the same year) do not mate for a year or so after the majority does. This is also true of polar bears.

In most areas of the Arctic, female polar bears breed for the first time at four years of age and give birth to their first cubs at the age of five.

Some of these young females will lose their first litters. Several will not mate until they are five or six years old.

There are also some interesting differences between populations in the age of first breeding of females. For example, in the lower central Canadian Arctic, adult females can have cubs for the first time at the age of five, but only about a third of them do. In both the Alaskan and Canadian Beaufort Sea areas, females do not have their first cubs until they are six.

A full year's difference in the age of first breeding between the polar bears of the Beaufort Sea and those of other areas is quite remarkable. The reason is not known, but it is probably influenced by the overall biological productivity of the different areas. For example, it is often stated that the waters of the eastern Arctic, such as Baffin Bay and Lancaster Sound, are much more biologically productive than is the Beaufort Sea. The waters of the Beaufort Sea come mostly from the depths of the polar basin, which is supposed to be one of the least productive marine water bodies in the world. If the overall level of biological productivity is lower in the Beaufort Sea, there may be fewer seals there than in other areas. If the seal populations are smaller, polar bears might catch them less frequently, and it might take a little longer for female bears to grow to physical maturity.

The majority of male bears reach sexual maturity around the age of six years, but that does not mean they will succeed in mating at that time. Several more years must pass for a bear to grow big, strong, and skillful enough to dominate his competitors. Most of the mating is probably done by males eight to ten years of age and older.

## How Often Do Females Have Cubs?

The rate at which a female can have cubs depends on how long it takes her to wean one litter and have another. In most areas, females keep their cubs with them for two and a half years. This means they can mate once every three years. Even though females in the Beaufort Sea mature a year later than do females in other areas, they keep their cubs for the same length of time. In contrast, about 40 percent of the mother bears of the western coast of Hudson Bay wean their cubs at only one and a half years of age. This means that many of the females in western Hudson Bay are capable of breeding every two years, which may enable them to have up to 30 percent more litters during their lifetimes.

Malcolm Ramsay investigated whether such early weaning meant a lower survival rate among Hudson Bay cubs, compared to those of other areas. He found there was no significant difference. Yearlings captured alone seemed to do as well as yearlings that remained with their mothers. Such a difference in productivity could have a significant impact on how the population might be managed.

We do not know why Hudson Bay females wean their cubs as yearlings, when mothers elsewhere keep them for a year longer. There are a few interesting possibilities. One idea concerns the minimum size that polar bears must reach to be able to hunt successfully. In our studies of the hunting behavior of polar bears in the Canadian High Arctic in the spring, we noted that polar bears must crash through hard snowdrifts to capture seals at their breathing holes and haul-out lairs (see also the chapter on behavior). Yearlings may be too small to hunt this way on their own because they are not heavy enough to break through the snowdrifts in winter. At the lower latitudes of western Hudson Bay, where it gets warm earlier, the snowdrifts may be softer and easier to break into. If so, it might allow a smaller bear to hunt successfully.

Another possibility is that Hudson Bay may be more biologically productive, thanks to the many rivers that discharge nutrients into it. This might

TABLE 2. **Litter Sizes of Polar Bears in Different Geographic Areas**

| Geographic Area | Approximate Latitude | Litter Size | | |
| | | Cubs of the Year | | Yearlings |
| | | Spring | Summer | |
| --- | --- | --- | --- | --- |
| Ontario coast of Hudson Bay | 55–57°N | 2.00 | 1.73 | — |
| Manitoba coast of Hudson Bay | 57–58°N | 2.00 | 1.60 | 1.68 |
| Southampton Island | 63–67°N | 1.94 | — | — |
| Southeastern Baffin Island | 62–67°N | 1.82 | — | 1.57 |
| Simpson Peninsula, Central Canadian Arctic | 68–69°N | 1.71* | — | — |
| Central Canadian Arctic | 70–73°N | 1.56 | — | 1.38 |
| Beaufort Sea | 70–74°N | 1.66 | — | 1.57 |
| Northeastern Greenland | 73–78°N | 1.71* | — | — |
| Wrangel Island, USSR | 71–72°N | 1.78 | — | — |
| Canadian High Arctic | 74–78°N | 1.63 | — | 1.70 |
| Svalbard | 77–80°N | 1.82* | 1.56 | 1.23 |

*wholly or largely based on counts before cubs left dens

result in larger seal populations. A greater number of seals could mean that bears of all ages might have less difficulty catching food.

A third explanation relates to the relative abundance of the different species of seals available to be hunted. Hudson Bay has a large population of bearded seals, which are four to five times larger than the ringed seals. A bearded seal kill means a much larger supply of carrion left on the ice. Since young bears get much of their food by scavenging the kills of older bears, this resource might be enough to help yearlings survive until they are large enough to kill seals on their own.

These untested ideas provide interesting areas for future researchers to investigate.

## Litter Size

The third major reproductive variable is litter size. Twins are most common in polar bears and account for about two-thirds of the litters in all areas. Single cubs are the next most common, accounting for 20 to 30 percent. The occurrence of triplets varies, being more common in lower latitudes. For example, of 119 litters caught in the spring on the Manitoba coast of Hudson Bay, 24 (20 percent) were cubs of the year, 79 (67 percent) were twins, 14 (12 percent) were triplets, and one litter had four cubs. In comparison, only 1 to 3 percent of the litters caught in spring on the southeastern coast of Baffin Island, Svalbard, the Canadian High Arctic, or the Beaufort Sea were triplets.

Litter sizes also vary because some of the cubs die. Quite often in triplet litters there are two normal sized cubs and a runt, which usually does not survive even to the following summer. Our nickname for these runts is "underbear," similar to an "underdog."

Simply because of natural mortality, one would expect the highest litter size to be recorded at the den before the family departs for the sea ice. As the cubs get older, some die for various reasons, and the average size of the litters decreases. Consequently, the mean litter size of yearling cubs will be smaller than that of cubs of the year. The average litter sizes of cubs as presented in Table 2 were recorded when the families were leaving the denning areas, or on the sea ice shortly there-after. They illustrate the variation in litter size in different areas.

Data on litter sizes of cubs in their dens are extremely difficult to obtain, but they do give some insight into the amount of mortality that occurs there. Most of what little information is available was collected by scientists accompanying Inuk hunters several years ago, when it was still legal to hunt polar bears in their dens. For example, from November 1952 to June 1956, eighty-one bears were killed by the Inuit of Pelly Bay in the Northwest Territories. Sixty-two of those were killed in their dens.

Although the sample sizes of litters from within dens may not look impressive, we should remember that these data sets took several years to collect and usually involved weeks of travel at a time by dog team, in bitterly cold temperatures in the early spring.

Father Van de Velde was a Catholic priest who lived and traveled extensively at all seasons with the Inuit from Pelly Bay. He was a remarkable man in any terms but he also had a life-long love of polar bears and kept invaluable notes on his observations of them. He recorded fifty-six litters of polar bears, most of which were still in their dens. The mean litter size was 1.71. When this value is compared with the mean litter size of 1.56 from cubs caught on the sea ice elsewhere in the Central Canadian Arctic in April, it appears there could be a mortality rate of 9 or 10 percent in the first few weeks after leaving the maternity den.

Richard Harington collected similar data from Inuk hunters during the early 1960s. On the basis of sixteen litters of cubs still in their dens on Southampton Island, he calculated a mean litter size of 1.94. On southeastern Baffin Island, about 800 kilometers (500 mi.) to the east at the same latitude, the mean litter size of cubs as they first reach the ice in April is 1.82, a difference of only 6 percent.

In his pioneering study of polar bears at Svalbard, the Norwegian biologist, Ødd Lønø, collected data from resident polar bear trappers from 1946 and 1967. He examined the ovaries collected from fourteen adult female polar bears for evidence of ovulation. He found that thirteen had released 2 eggs and one released 3 for an average of 2.07. During the same period, he obtained data from twenty-four litters of cubs that had an average litter size of 1.67. Even though his sample sizes were not large, it is clear that a certain amount of egg loss, prebirth mortality, or postbirth mortality was taking place in the den.

Of course, there is no way of knowing if a female with no cubs might have lost her whole litter unless you know her history. Malcolm Ramsay calculated loss of cubs using tagged animals with known histories from the western coast of Hudson Bay. From the time they left their dens in the spring and when they returned to shore again in late summer, families experienced a cub loss of 38 percent.

The largest average litter sizes of polar bears anywhere in the world were recorded along the western and southwestern coastlines of Hudson Bay in the 1970s. Each spring from 1970 through 1976, Dale Cross, Dick Robertson, and others from the Manitoba Department of Natural Resources counted the tracks of polar bear families as they plodded through the black spruce forests out to the coast. Their tracks are easy to see from the air, so by doing a series of flights to count new sets of fresh tracks after snowstorms, a total count can be made. Tracks can be followed backward to find the actual den sites. The number of cubs born can be estimated by summing the number of tracks and the number of cubs seen.

Between 1974 and 1978, George Kolenosky and Paul Prevett did a similar survey for the Ontario Ministry of Natural Resources along the northern Ontario coast. In both studies, the average litter size was 2.0. A greater frequency of triplets in these areas yielded the high average litter size.

Another intriguing aspect of litter size is that it

Females normally nurse in a sitting position. Here she feeds two nine-month-old cubs. After suckling, cubs usually fall into a contented sleep.

Subadult bear licking its paw to keep it clean

Young cubs walk close behind their mothers.

The Tundra Buggy was sitting still while people looked for bears. A female and her yearling cub passed nearby, and the cub approached to about 10 meters.

The female was nervous about the vehicle and tried to push her cub from behind to get him away from the vehicle.

The female then walked away, turned, and sat up.

She waved her paws about to attract the attention of her cub.

She gave calls to the cub that sounded like braying or moaning.

She finally enticed him back to her and led him away a short distance. Once away from the Tundra Buggy, she sat down and allowed him to nurse.

When the cub was finished, he turned to take one last look.

He then followed his mother away from the area. (Photographs © Carol Dixon and Fred Treul.)

Two large adult males meet and make facial contact as they prepare to meet in ritualized play fighting. Although the encounters are rough, and may last up to forty-five minutes, injuries are rare.

One male stands to try for the advantage in striking.

They grapple and push as each tries to knock the other off balance.

One is pinned down but he frees himself to attack again as they stand to bite and swing at each other with their forepaws.

After a brief locking of the jaws, the loser walks away.

Typical postures in play fighting by adult males. They stand, sometimes singly and sometimes together, as they seek a position of advantage from which to attack one another.

Occasionally they appear like dancers when each bites the other's neck and holds with the forepaws.

Each tries to knock the other off balance and pin him down if possible.

A young adult sits up to watch, then straightens his hind legs to stand.

can change in the same area over time, possibly as a result of ecological changes. For example, in the Beaufort Sea in the 1970s, the mean litter size of cubs of the year in the spring was 1.66 (Table 2). However, when we went back in the mid-1980s, we found the mean litter size was 1.84 although the reason for this is not apparent. Similarly, in the 1980s, in western Hudson Bay, we found a mean litter size of 1.94, a small drop from 2.0 in the 1970s. This tells us that an ecosystem is not a constant thing and that the bears are constantly adjusting their reproductive effort in response to change.

## More about Cubs

An examination of the sex ratio of multicub litters indicates that most are fraternal rather than identical twins. Several factors can influence their weight, including the size of the litter. In the western Hudson Bay population, Malcolm Ramsay found that single cubs in the spring weighed about 30 percent more when they were leaving their dens than did cubs from litters of two. In turn, cubs from litters of two weighed about 30 percent more than did those from litters of three.

In litters of three, one of the cubs is often considerably smaller than the other two. We sometimes jokingly refer to these as two-and-a-half-cub litters. Although we tag the "underbears," we almost never see them again. They probably cannot compete with their siblings for milk from the mother, or parts of the seals she kills. As in other mammals, larger sized young have a better chance of surviving.

## The Reproductive History of a Special Bear

I cannot overemphasize the value of collecting data on the same individual polar bear over many years, in what are called "longitudinal histories." By doing this, we get some insight into long-term trends and fluctuations that are missed in projects that last only a year or two. Nowhere is that more important than when studying reproduction. We have long-term information on quite a few female polar bears but none is as valuable as a bear from Churchill who became known as Linda.

Normally we do not give the bears names. For one thing, it might become confusing just trying to remember them all. We have tagged over 1,800 individuals just in the population where Linda came from, on the western coast of Hudson Bay. More important, though, is that sometimes when scientists name animals, there is a risk that they will start to see the attributes of the namesake in the animal. How would you react to bears called Warrior or Wimp, for example? In fact, Linda is the only polar bear we have given a name to. And, although I do not normally like naming research animals, she was so remarkable that she earned and kept her distinction. In the computer however, she is just X505. All the tagging of polar bears worldwide is coordinated through the Polar Bear Specialists Group of the IUCN to ensure that the same number will not inadvertently be used by two countries. The X denotes that a bear is tagged in Canada, an A for Alaska, N for Norway, and so on.

Linda was born around New Year's day, 1965, in a snow den somewhere south of Churchill. In late February or early March, she left the comfort of her den to follow her mother through the snow to the sea ice. We do not know if Linda had any brothers or sisters, but since two-thirds of all litters consist of two cubs, she probably had a sibling.

Linda became X505 on 13 October 1966, when she was first captured and tagged as a yearling by herself near the rocket-launching range a few miles east of Churchill. She had come in to snoop around the garbage there and soon found herself in a foot snare. The following fall, Linda was caught again near the Churchill dump and fitted with a radio collar, which she promptly got rid of. No further information was gathered from her that year. In 1968, she was caught at the dump again and had another radio collar put on. This time she kept it for a few months after freeze-up and was tracked far out onto the bay to the northeast of Churchill. This gave us some of the first information on where the famous bears of Churchill go for the winter.

Most female polar bears in western Hudson Bay mate for the first time in the spring of their fourth year. Unfortunately, we do not know how old Linda was when she had her first cubs because she was not captured again until 1972 when she was seven. At that time, she had a single cub with her, although likely it was not her first. She and her cub were captured near the dump and then released at a place called Bird Cove about 14 kilometers (8 mi.) along the coast east of town. Her cub came back to the dump the following year and was captured in foot snares eight times through the fall. That cub has not been seen since but the fact she was known to have been a lone yearling in 1973 strongly suggests that her mother was on a two-year breeding cycle, was pregnant again, and had departed into the maternity denning area. We cannot confirm that, however, because she was not seen in 1974 when she should have had cubs with her. Another two years later, though, in 1976, she had her first recorded set of triplets, which would be right on schedule if she was on a two-year breeding cycle and had cubs in 1974 as we suspect she did. This time, Linda and all her cubs were flown by aircraft about 240 kilometers (150 mi.) in a straight line southeast of Churchill (double the distance along the coast) to keep them out of the dump. Two of her cubs from that litter have not been seen since, but one was back the following year, again as a lone yearling, suggesting that Linda, still on her two-year cycle, was back in the denning area getting ready to

have cubs again in 1978. As in 1974, she was not seen in 1978, but another two years later, in March of 1980, she was captured with three more cubs as she left the denning area. By October of the same year, the whole family was at the dump in Churchill again. This time, the family was not moved but was observed all fall as part of a study to determine if there was any significant ecological benefit being gained by the bears that fed in the dump compared to those that did not feed there. She received her third radio collar that fall and continued to give valuable information on movements when she returned to the ice of Hudson Bay after freeze-up.

Two years later, in the fall of 1982, now seventeen years old, she was back at Churchill again with two more cubs. By this time, the Manitoba Department of Natural Resources had built the "Polar Bear Jail" for the purpose of locking up bears that came around town and might cause a problem. The reason for incarcerating them was to remove the reward they might get by coming to the dump. The bears are kept in jail until freeze-up and then returned to the ice where they can once again feed on their natural food, seals. Since the bears have enough stored fat to last until freeze-up, they are not fed. Although expensive, this admirable program has saved many individuals from becoming "problem bears" and perhaps being shot.

Linda and her cubs were among the first inmates of the polar bear jail. While there, Linda helped out science again by wearing a mock-up prototype of a satellite radio collar for a month to test the practicality of the design.

She was released back onto the ice after freeze-up but this time, instead of remaining on her characteristic two-year breeding cycle, she returned to Churchill in the fall of 1983, accompanied for the first time by yearling cubs. As old age began to catch up with Linda, the size of her litters became smaller and the intervals between them got longer.

Sometime during her period ashore that year she also met up with a porcupine and got a mouthful of quills, which we removed with loving care. It seems that polar bears, unlike most land predators, have been separated from porcupines for so long they no longer recognize them as dangerous. Along the coast of Hudson Bay, the ranges of these two mammals still overlap and almost every year we find at least one bear with quills in its mouth. I often wonder what happened to the porcupine!

This time, the family was held for half the winter and was not released onto the sea ice until February 1984. Apparently that did not slow Linda down, because in the fall of 1985, at the ripe old age of twenty, she was back at the Churchill dump with two more cubs and promptly landed back in the polar bear jail. This time she helped out science by participating in some feeding experiments designed to test the physiological adaptations of polar bears to fasting.

That was when the debate about Linda's future began. The management biologists and conservation officers responsible for keeping the town safe from problem bears pointed out that although Linda herself had never been reported in town, several of her cubs had caused damage there and two had to be destroyed. Their suggestion was to send Linda to a zoo to see out her days where she would not contribute more potential problem bears to Churchill. The research biologists wanted Linda released because she was felt to be nearing the end of her reproductive life and it was unique to have so much information on a single bear. Furthermore, it did not cost anything to find out how many cubs she had because she always brought them in to show us.

In any case, we thought it unlikely that she would have more than one more litter. We had already tagged eleven of her cubs and likely there were another six or so we had not recorded. However, it was finally decided that the zoo in Albuquerque, New Mexico, would be home for Linda, and so ended the wild chapter of the history of a truly remarkable bear (p. 173). No wild polar bear has been better known for so many years or through so many litters, radio collars, and other scientific experiments. Maybe none will ever be again. Yet, for all that, Linda remained one of the gentlest and most good-natured polar bears any of us have ever known. She also achieved some accidental fame in the early 1980s when she appeared on the Sierra Club's Christmas card: no one noticed the tag in her ear when the photo was printed but we recognized her instantly!

Of the eleven cubs we tagged, four are known to have died and three others have gone to zoos. Four have not been seen since they were released. We do not know what befell them or if we can still look forward to capturing them and reviving Linda's legend. Unfortunately, there is no way to recognize the cubs we think she had but did not tag. Maybe that is best. It is nice to think that out there somewhere, her descendants are still roaming over the ice fields while their mother spends her remaining years where she is admired and appreciated every day by her countless fans.

# Behavior

Our understanding of the behavior of any animal comes from many sources of information. Sometimes you can watch an animal directly, but often impressions are drawn from more indirect sources such as tracks, the remains of kills, analysis of scats, or by analogy from other anecdotal observations. In the case of a solitary carnivore like the polar bear that lives at low densities over a habitat as vast and relatively inaccessible as the sea ice of the Arctic, much of our information comes from the indirect method.

Over the years, we have caught several thousand polar bears all over the Arctic. We have caught them in different seasons, habitats, and geographic areas. Many bears are found by following their tracks, often for 100 or more kilometers (60 mi.). When tracking, your eyes are glued to the ice in front of the helicopter and you note every turn the bear makes, every ridge he walks over or around, and the areas he walks through without stopping. He follows refrozen cracks in the ice because of the seal breathing holes his past experience tells him are there. Periodically you lose the tracks in an area where the wind has covered them with drifting snow, or they get mixed up with the tracks of other bears, or they just disappear for no apparent reason. Where did

he go and why? Sometimes you find the bear again and sometimes you do not. But whether you realize it or not, in the process of deciding where to look to relocate them, you are applying your knowledge of their behavior. You see where bears have tried to catch seals but failed. Blood stains and remains on the ice tell where a meal was consumed. From such observations, you begin to form ideas about what kinds of ice habitat are best for hunting seals. The intervals between rests, periods of nursing by cubs, or defecations are all recorded in the ephemeral format of snow on the sea ice. Sometimes you see tracks or signs of activity you have no idea how to interpret.

Occasionally, you see bears before they see you, and you have a brief opportunity to watch one doing something. These brief impressions are often difficult to interpret. Worse, they may lead to misleading ideas.

The frustrating part of gathering information in this manner is that there are not enough facts to permit meaningful conclusions. However, it is this backdrop of lore, accumulated through personal observations and the reports of Inuit hunters, explorers, and other biologists, that forms the basis for much of our understanding of the behavior of wild polar bears.

## Watching Undisturbed Wild Polar Bears

On 24 July 1973, I left Resolute for Radstock Bay on the southwest coast of Devon Island to see if it was practical to study the behavior of polar bears on their natural habitat, the sea ice. Caswall Tower is a 200 meter (640 ft.) high rock tower like a mesa, 10 kilometers (6 mi.) along the west coast from the mouth of the bay, which affords an almost unbelievable view of the sea ice (see Fig. 5). Chuck Jonkel, formerly of the Canadian Wildlife Service and now at the University of Montana, first told me about the place. He had an Inuk in Resolute make a small plywood hut to put on the top to use for a camp.

During the summer, Lancaster Sound turns into open water so that the only remaining ice in the area is in the bays along the southern coast of Devon Island. This has the effect of concentrating the bears that want to continue to hunt for an extra few weeks before open water takes over those bays as well. I thought this might enable me to carry out extended observations of the behavior of polar bears gathered there. However, most of my colleagues were skeptical of the idea, so I took my wife, Stella, with me as a field assistant and billed the trip as a pilot project only.

The first morning was overcast, with fog banks looming by the edge of the ice at the mouth of the bay. We scanned the ice with binoculars and telescopes for some time without seeing any bears, although seals, fulmars, guillemots, and gulls were abundant. After lunch, we decided to go for a walk to see if there were any tracks or scats on the beach that might indicate the presence of bears in the area. We had only been walking about an hour and a half along the eskers and meadows on the flats near the coast when Stella noticed that a white rock a couple of kilometers away was moving. Sure enough, it was a large adult male walking across the land headed in the general direction of the ice near the base of Caswall Tower. We climbed back up the scree slopes to the top of the tower, set up a telescope, and began to watch. The bear ambled over to the beach, wandered along the edge of the ice, and finally went to sleep on a gravel ridge. The fog rolled in and the bear disappeared from sight. When the fog lifted the following afternoon, the bear was gone but we were greatly encouraged and soon spotted another, this time out on the ice. The weather then cleared and for the next two weeks, we watched bears continuously with the aid of the twenty-four-hour daylight that prevails at that time of year. With the telescopes, we could watch bears up to about 7 kilometers (3 mi.) away comfortably and up to about 11 kilometers (6 mi.) if the light was particularly good.

Sometimes it was too cold to sit still outside and observe. The uninsulated shack was not much better. There were only a few tiny windows and they were too high for convenience. We put a chair on top of the camp table and left a Coleman stove on the floor for heat. This let us look through the clear plastic windows with our telescopes. By the time the ice broke out of the bay and the bears had disappeared, we were exhausted. But we had collected 603 hours of observations, the most that had ever been done up to that time, and we had learned an immense amount about the behavior of undisturbed polar bears.

After that summer, others became involved on the behavior project. We quickly insulated the shack on Caswall Tower and put in proper windows so you can use a telescope while sitting down. When we began our winter observations, we found that most of the activity took place on the rough ice near the mouth of the bay and along the south coast of Devon Island. To watch this area, we built a second observation hut at Cae Liddon at the southwest entrance to the bay. Since then, we have accumulated several thousand hours of continuous observations of bears of all ages and sex classes.

Unmarked bears were anonymous. They

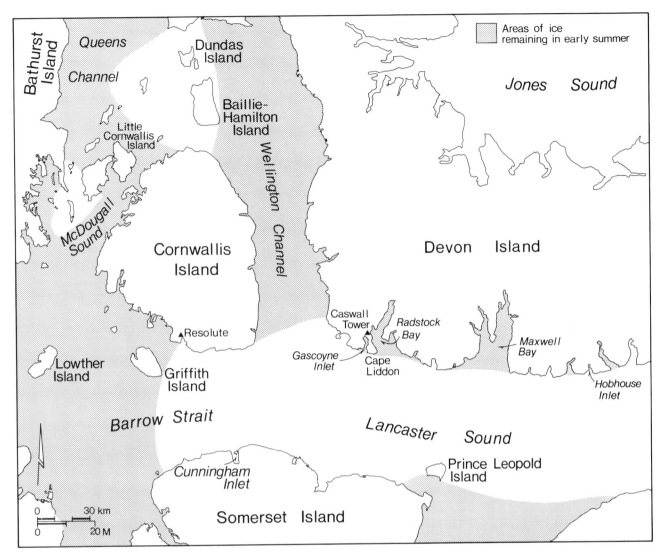

Fig. 5. Lancaster Sound–Barrow Strait area

yielded a few hours, or even days, of information before disappearing. Others were captured at the beginning of a trip and had numbers painted on them. The age, sex, and size of these animals were known and they were followed off and on, sometimes for a few weeks, resulting in some of the most valuable information of all. More than anything else, the behavior project has influenced my understanding of the bears themselves and how the arctic marine mammals have evolved with each other over the millennia.

Nothing quite equals the sense of total fascination and respect that overcomes me when I spend several hours watching undisturbed animals hunting on the sea ice. Everything from the kaleidoscope of several years of experiences comes together. No longer are the bears just interesting specimens, scientifically identified by ear tags. They become the ultimate arctic animal, entirely adapted for life on the sea ice, an integral part of the ecosystem.

One of the first things to strike you is that polar bears rarely do things quickly. Their steady lumbering gait averages about 5.5 kilometers per hour (3.5 MPH) and it can continue for hours at a time. Their great heads swing gently from side to side as they walk, making them appear mis-leadingly inattentive to their surroundings. Peri-odically, they stop to look around or raise the nose to test the wind before plodding on. Yet there is little that passes unnoticed. Their eye-sight is about the same as that of a man but it is their prodigious sense of smell that they rely on for information. It is scent, like an invisible code, that identifies a place to hunt on the ice, tells them if other bears are upwind or if there is a seal carcass to scavenge upon, or just notes some-thing interesting that warrants investigation. As you continue to watch, you realize there is abso-lutely nothing casual about these animals. Their entire existence is built around hunting and the conservation of energy. Their actions, if you have the patience to watch, tell you what it means to be a polar bear.

## Seal Hunting in the Summer

Polar bears hunt seals in two different ways, by stalking and still-hunting. There are several vari-ations on each approach. The best-known meth-od, and certainly the most exciting to watch, is stalking. This begins when a bear notices a seal lying on the ice, usually at a distance of a few hundred meters. Instantly the lackadaisical am-bling is gone and the bear suddenly freezes, sometimes in mid-step. It stands motionless, and peers intently at the seal. Sometimes it will stand for several minutes, apparently evaluating how to get close enough to make a final charge. Usually, the bear lowers its head and then walks slowly and steadily toward the seal, sometimes in a semicrouched position as it gets closer.

Ringed seals are quite alert on the ice and con-stantly raise their heads to look around for preda-

tors (p. 127). When this happens, Inuk seal hunt-ers are motionless until the seal puts its head down again. Curiously, the polar bear is not. It just keeps steadily moving toward the seal. When the bear gets to about 15 to 30 meters from the seal it suddenly charges, at quite incredible speed, while the horrified ringed seal attempts to get into its hole. Ringed seals usually haul out alone at a breathing hole. It is easy to see why. If there are more than two seals by a single breathing hole, they get in each other's way when trying to escape a charging bear. This gives the polar bear the split-second advantage it needs to get its claws or teeth into one of them. After grabbing a seal, the bear bites its head many times rapidly before dragging the carcass away from the hole to begin eating it.

Some bears specialize in what I have called the aquatic stalk. Like the bears that stalk seals across the ice, they seem to memorize the route to the seal before they slip into the water. There are two kinds of aquatic stalk. In one kind, the bears dive and swim underwater between holes in the ice. They surface to breathe so stealthily that only the tip of the nose breaks the water be-tween dives as the bear moves closer to the seal. The longest dive I timed between breathing holes was 72 seconds. Finally, it gets to the last avail-able breathing hole before reaching the seal and slips out of sight again. After an eternity of sus-pense-filled seconds, the water in front of the seal explodes as the bear suddenly claws its way onto the ice after its prey. Despite this imaginative ap-proach, more often than not the seal manages to evade the lethal claws and reach the water where it can outswim the bear. In one instance, a bear stalked a bearded seal from a distance of over 300 meters (300 yds.) and missed it by less than a me-ter. The seal may have seen him coming up through the water because it was already diving off the ice as the bear surfaced.

I sometimes think about this type of hunting in relation to a question I was asked by an Inuk hunter in northern Quebec. He wanted to know if I had ever seen a sea bear. When I asked what they looked like, he said they appeared the same as any other polar bear but they were such good hunters that they grew too big and fat for their legs to still be able to support them as they walked on the ice. Thus, they had to swim all the time, hunting along the edges of leads. There, they can surface suddenly and seize an unsuspecting seal with their long forepaws. He told me they were very dangerous and this was a good reason not to camp right beside the floe edge. With memories of many aquatic stalks in my mind, I assured him I would not.

Another type of aquatic stalk involves the use of the water-filled channels that form on top of the sea ice in the summer. The channels, and the larger pools they connect, are about 25 to 50 centimeters (1–1.5 ft.) deep. Again, the bear stands and studies the route to its intended dinner before lying flat in the channel and sliding along almost completely out of sight (p. 126).

Occasionally, bears use channels that pass so close to the base of the cliff below our camp that we can clearly see the black snout and small dark eyes, barely above the surface of the water, as the bear silently slides along. It propels itself with its paws, extended out to the side rather than underneath, in order to allow the animal's profile to remain below the level of the ice.

In 1973, one adult female had a hilarious individual modification to this approach. She lay with the front half of her body flat in the water, but kept her rear legs straight so that her whole hind quarters towered over the ice. From a distance, it looked like the seal was being stalked by an iceberg! Curiously, this slow-moving white object did not appear to unduly alarm the seals, even when it got quite close. Several years later, I saw another bear doing this. I wondered if it was the same female, or if she had taught the trick to one of her cubs.

There was one particular aquatic stalk that will

always stand out in my mind because of what it shows about the ability of a polar bear to follow a plan. It took place during the summer, about a kilometer from our camp on top of Caswall Tower. With the telescope, we had a clear view of what happened. Like most other bears, this female plodded across the ice seemingly unaware of her surroundings. Then she spotted the unmistakable black outline of a seal lying on the ice. She froze instantly and, for several minutes, peered intently at the series of channels in the ice that led toward the seal. Then she lowered herself into the water and glided silently off in the direction of her quarry. About halfway to the seal, she came to a fork in the channel. The left-hand channel headed toward the seal while the one on the right slowly veered away. She paused for a moment and started up the wrong channel for about a body length, and then stopped. Without lifting her head to check her bearings, she slowly backed up and headed off again along the correct channel. I was amazed. I have seen several bears lift their heads slightly to check their bearings but I have never seen such a demonstration of conscious memory. It reminded me of my old friend Jimmy Memorana, an experienced Inuk polar bear hunter from Holman. "Keep looking around when you are working on the sea ice," he once told me in his usual patient manner. "If the bear is hunting, you won't see him until he comes for you."

One of the most recurrent legends about a hunting polar bear is that it will cover its giveaway black nose with its paw when stalking a seal. None of the Inuk hunters who have mentioned this to me have actually seen it themselves but the story exists in the oral history of several areas. Peter Freuchen, the famous Danish explorer (and storyteller!) claimed to have seen a bear do this, as did the arctic physiologist Kare Rodahl on the coast of East Greenland in 1939–40. Perhaps. All I can say is that in several thousand hours of watching polar bears hunt, none of us has ever seen a polar bear cover its nose. If it does happen, it certainly is not very common.

In several areas, the sea ice melts completely in late summer, so the bears have to wait on land for the sea to refreeze. In most places, it seems that no hunting of seals goes on at this time. There is one very interesting exception. In August 1978, Don Furnell, a biologist from the Northwest Territories Wildlife Service, and David Oolooyuk, an Inuk hunter, were watching a large adult male polar bear swimming in relatively shallow water. About 75 meters away, a ringed seal appeared to be diving to feed and surfacing at about the same location each time it needed to breathe. The bear swam toward where the seal was breathing. When the seal surfaced, the bear lay motionless in the water. Finally the seal came up only half a meter away and the bear lunged, biting it firmly in the back. They speculated that the seal might have surfaced by the bear because it thought he was just a piece of ice. This is plausible. It is well known to Inuk hunters that seals are often found around pieces of ice during the open water season. They did not know if more than one bear was hunting this way but other fresh seal carcasses were found along the beach. Bears are quick to learn from both experience and observation. Possibly this specialized behavior was practiced only by a relatively small number of bears in a fairly local area.

When still-hunting, bears remain motionless beside a breathing hole or at the edge of a lead and wait for the seal to surface to breathe. In the summer, about 80 percent of the hunting is done this way. Only recently have we recognized this as the main method of hunting. Early explorers and naturalists who wrote the first anecdotal accounts of polar bear life history may have thought a bear lying still-hunting was just sleeping. Certainly, the spectacular nature of the stalking hunts captures the imagination more.

As the bears walk across the ice in search of seals, they stop to sniff at breathing holes and along the edges of cracks in the ice. Individual seals often haul out at the same spot for several days in succession if they are not bothered. When a bear encounters one of these places, it likely detects the scent on the ice and recognizes it as a place where a seal may surface to breathe again.

The most usual position for still-hunting is to lie on the stomach and chest with the chin close to the edge of the ice. Sometimes bears prefer to sit or stand (p. 126). When hunting at breathing holes, it is essential to be absolutely silent. Seals are easily frightened and will go somewhere else to breathe if they hear something at the surface. I have tested the transmission of noise through the snow-covered ice myself by putting a hydrophone in the water through a seal breathing hole and then moving. Even the slightest movement of my mukluk on the snow transmits a loud crunching noise into the water. The sound of a person walking can be heard clearly up to 400 meters (1,300 ft.) away. No wonder a polar bear must remain so still!

Most still-hunts are less than an hour long but some may go on for much longer. That is probably why most bears prefer to save energy by lying down in a comfortable position. Another possible advantage to lying down is that it presents the lowest silhouette against the sky (when viewed from below by a seal).

When a seal surfaces to breathe, the scene of peace and tranquility is instantaneously galvanized into action. In a single movement, the bear bites the seal on the head or upper body and flips it out on the ice, wriggling much like a trout that has just been pulled out of a stream. The bear then bites it many times about the head and neck and drags it several meters away from the edge of the water before starting to eat it.

## Seal Hunting in Winter

During the cold weather, since few seals haul out on the ice, almost all of the hunting is done by still-hunting. Wherever there is open water, seals will surface to breathe. The seals breathe at cracks in the ice for as long as possible because it is easier and it doesn't restrict their movements quite as much. After the ice is completely frozen, they must keep their breathing holes open themselves for the rest of the winter. When the ice is young and thin, the seals just push up through it with their heads. These holes are easily accessible to the bears. As the winter progresses, and the remaining cracks freeze, the seals' movements are restricted to the area around the breathing holes they maintain themselves (p. 126). In time, drifting snow accumulates over the cracks and pressure ridges, removing all sign of the location of the breathing holes. Later in the winter, when the drifts over the breathing holes deepen, some ringed seals excavate small haul-out lairs, like little snow caves, above their breathing holes. In the spring, the pregnant females use these subnivean (beneath the snow) haul-out lairs to give birth to their pups, out of sight of the ever-present polar bears.

The acuity of a polar bear's eyesight is probably not much different from that of a human being but its sense of smell is superb, and that is what it relies on most for information about the world around it. At no time is this more obvious than when a bear is hunting for the breathing holes and haul-out lairs of ringed seals beneath the snowdrifts.

In the winter and early spring, the cold white surface of the arctic sea ice appears as lifeless as any desert one might imagine. Snowdrifts accumulate along the edges of the long pressure ridges of jagged ice that zigzag across the horizon. No seals are to be seen. Only the occasional track of a polar bear or an arctic fox reminds you of the presence of animals. Yet life is abundant if you know where to look. For a bear, though, looking is less important than smelling. His extraordinary sense of smell enables him to locate ringed seal breathing holes beneath the snow.

In the spring, polar bears especially seek out the seal birth lairs with pups in them. The pups weigh only about 5 kilograms (10 lb.) at birth but they grow quickly to about 20 to 25 kilograms (40 to 50 lb.) by about six weeks of age when they are weaned. At that time, they can be up to 65 to 75 percent fat, which is the part the polar bears like to eat most.

As a polar bear plods steadily across the ice, it appears oblivious to the possibility of anything else being about. Yet it constantly sniffs the wind as it walks. Occasionally, it stops to look about or check again for some scent it might have just caught briefly.

When a change in a bear's behavior indicates that it is going to investigate something under the snow, it is usually only about 50 to 100 meters (164 to 328 ft.) from the point of interest. At first I thought that this distance indicated how far away a bear could detect a breathing hole under the snow. Then I was fortunate enough to do some studies on ringed seal birth lairs with the arctic seal biologist Tom Smith and his incredible dog, Bug. Along with Jimmy Memorana, who is a walking encyclopedia of life on the sea ice, we wanted to study the different types of ice and snowdrifts used by seals. This, we thought, would help us to better understand the hunting behavior of polar bears. At Jimmy's suggestion, we used a dead seal pup to train Bug to sniff out seals, " . . . like the old people used to do long ago." Bug was a trained Labrador retriever, and at the cry of "Go find those birds!" she tore off across the sea ice sniffing out seals all over the place. We were astonished at how many she found so quickly, often in relatively small areas. But we were even more impressed at how far away she could smell things. Bug could detect a seal breathing hole, under a meter (3 ft.) of snow, from a measured distance of a kilometer or more!

I am sure that a polar bear's sense of smell is as good as a Labrador retriever's. This means that when a bear tests the wind, it must be checking a veritable smorgasbord of smells. Clearly, the bears are being very selective in deciding where they are going to hunt, although we do not know much about what their cues are.

In the early spring, most of the still-hunting is done standing rather than lying down. Once a bear decides to hunt, it creeps up slowly and carefully to the chosen spot. Then it stands absolutely still with its front feet and hind feet closer to each other than would be usual if it were standing normally. This slightly odd-looking posture seems to allow the bear to shift its weight back onto its hind feet without moving them, which would transmit sounds through the snow to a seal below. When the bear hears or smells a seal beneath the snow, it slowly stands up on its hind legs, raises its front limbs, and crashes down with its front paws in an attempt to break through the roof of the snow cave. Sometimes a bear is successful on its first try, but we have often seen large adult males and females that needed to jump and dig frantically several times before penetrating the hard surface of the snow. By this time of course, the seal has gone.

The Inuk hunters of long ago must have watched polar bears hunting seals in this way, because their methods were a direct copy. The journals of explorers such as Francis Hall contain vivid accounts of hunters standing motionless with a spear for many hours in freezing temperatures, waiting for a seal. It is also likely that watching polar bears stalk seals across open expanses of ice, without being noticed, first gave Inuk hunters the idea of making a screen of white cloth to hold in front of themselves. There is no way of knowing if this is true, but it is certainly plausible.

When polar bears break into seal lairs, they do so right over the breathing hole. Sometimes, if a bear catches a very young pup with little body fat, it will kill it but not eat it (p. 127). Occasionally, if the birth lair is in a deep snowdrift, the bear may lie head-first down the hole he dug into the lair. This prevents the sunlight from shining onto the breathing hole. Inuk hunters have told me this makes the ever-cautious ringed seal mother think its lair is still intact and safe to return to. In any case, the female sometimes comes back to check on her pup . . . and gets caught. Nursing female seals are very fat and provide much more nutrition for a polar bear than do their pups.

Again, the Inuit of earlier times may have learned a seal hunting method from the polar bear. I had the opportunity a few years ago to watch some of this now-disappearing form of hunting practiced by an expert. I was working along the coast of southeastern Baffin Island with Tom Smith again, and an Inuk hunter named Ipeelie Inookie. It was late in the spring and warm, so that one could occasionally detect a slight depression in a snowdrift, indicating the sagging roof of a seal birth lair beneath. Ipeelie approached the drift slowly like a stalking polar bear and then took a short run and a jump. He broke through the roof of the lair with his feet, and then dove in headfirst to try and seize the pup. Then he would dangle the pup in the water on a rope in case the mother came back to check on it. Sometimes she did, which was a testimonial to this Inuk and polar bear method of hunting.

## Learning to Hunt

During the two and a half years that polar bear cubs stay with their mother they watch her hunt thousands of times in a wide variety of seasons and circumstances. Like most predators, they learn much by watching their mother and then copying her behavior. During the summer, cubs of the year do almost no hunting. Instead, they follow closely behind their mother, watching everything she does, and sniffing in all the same places. They recognize the instant change in her behavior when she begins a stalk or lies down to still-hunt. That is the signal for them to lie down immediately and not move until she is finished. Usually the cubs are quite well behaved and wait patiently, but some will approach before she is finished. The females often respond to interruption with a short aggressive charge and sometimes even a cuff that bowls the cub over. After that the cub is much more patient again.

The intense interest shown by one cub was really quite comical. Its mother was a master of the aquatic stalk. She would flatten herself into a water channel in the ice and then painstakingly push herself toward the seal. Meanwhile her cub stalked along a few feet behind her, in plain view, watching its mother intently. Needless to say, the seals went down their holes long before they were in any danger.

When there are two cubs it is not uncommon for them to get distracted after waiting awhile and begin to play. One cub initiates things by biting or pushing its sibling who, of course, promptly responds. Soon they are biting, rolling over, and chasing each other back and forth all across the ice. In the summer, they seem to particularly like to run through shallow pools.

Sometimes single cubs get tired of waiting as well. One yearling cub I watched was particularly energetic. He amused himself by making long running leaps headfirst into melt pools in the ice, resulting in some quite spectacular splashes. As he was flying through the air one time, a seal popped up in the water right in front of him. The yearling grabbed it by the head, pulled it out on the ice, and killed it. But instead of beginning to eat his kill immediately like most bears, he raced about on the ice with the carcass, throwing it up in the air. Then he began to throw it into pools of water, dive in, and retrieve it. Finally the yearling's mother looked over, saw the seal, and raced over to begin eating it immediately. From the small size of the seal, it was only a few months old and not very wise about predators. No adult seal would have surfaced anywhere near where

the cub was galloping around so energetically. The abundance of these fat but naive young seals in the late spring and summer is extremely important to young bears as they learn to hunt for themselves.

Yearling and two-year-old cubs hunt about 4 and 7 percent of the time respectively, while their mothers spend from 35 to 50 percent of their time hunting. The yearlings we observed caught only one seal per twenty-two days of hunting, while their mothers caught one every four to five days. Even though two-year-olds didn't hunt very much, they caught an average of one seal every five to six days, indicating they had learned a great deal more in the additional year of tutelage.

Another factor that probably influences the number of seals killed by cubs of different ages is the way in which they choose places to hunt. Cubs of the year and yearlings tend to stay close behind their mother, watch her movements closely, and imitate them. If these younger cubs hunt when the female lies down to still-hunt, they do so more or less where they were standing when the female stopped moving. So, there is little independence of choice involved in picking a hunting site. Two-year-old cubs also follow their mother but may range up to 1 to 2 kilometers away from her. They choose their hunting sites more independently.

By the time a cub is two and a half years old, it has learned to hunt reasonably well but is still not motivated to spend a large amount of time practicing, possibly because its mother is still supplying it with food. More simply put, it is probably just lazy. During the early spring, yearlings and two-year-olds do almost no hunting. They may just be too small to break through a hard snowdrift fast enough to catch a seal. Part of the reason that polar bear cubs stay with their mothers for so long may simply be that they need extra time to grow big enough to be able to hunt successfully during the winter.

The average length of lying still-hunts of adult males was about 65 to 70 minutes. That of adult females with cubs of all ages was only about 35 to 40 minutes. The average hunting time of the females was probably shorter because of interruptions by cubs. Even so, the adult females still caught seals more often than males, in spite of the fact that the two groups spent similar proportions of their total time hunting.

## Feeding Behavior

Immediately after a polar bear captures a seal, he drags it from the edge of the water. He bites it several times about the head and the neck to make sure it is dead and often drags it several more meters from the water. He then begins to eat the seal immediately. He holds down the carcass with a forepaw, bites into the body, and rips upwards. At first, pieces are ripped off and swallowed with a minimum of chewing. This rapid and voracious initial feeding is important, because when a bear kills a seal, any other bear downwind will soon catch its scent and come to investigate. Adult males are particularly quick to approach the kill of another bear as the chance is good of being able to take it away and get a free meal.

If the kill is a ringed seal, the skin and fat are eaten first and sometimes in a very exacting manner. The fat is the preferred part of the seal. As the meal progresses, one can sometimes see the bear delicately using its incisor teeth, almost like shears, to snip away the fat without taking the meat. The skin of a bearded seal is thick and apparently not as appetizing to most bears. Often, the skin is peeled away from the fat but not much of it is eaten. Near the west coast of Banks Island, I once found the complete skin of a bearded seal, which had been removed from the carcass and placed hair side down on the ice. All the fat had been shaved from it as neatly as if a hunter's wife had done it with her *ulu* (a special skinning knife, with the handle positioned like a T above the middle of its crescent-shaped blade, used only by women).

Washing is an integral part of polar bear feeding behavior. In the summer, after an initial feeding period of 20 to 30 minutes, a bear would usually go to a pool of water. There it alternated between rinsing and licking its paws and muzzle. Bears we watched washed for up to 15 minutes after finishing feeding. In the winter, when water isn't available, a bear substitutes rubbing its head in the snow and rolling on its back for washing. Even at the Churchill dump, where some of the white bears get so dirty they look more like black bears, they periodically walk a kilometer to the bay. There they swim until they are relatively clean again. We have handled several thousand polar bears over the years and found very few to be dirty. When one considers how greasy seal blubber is, that is quite a testimony to how clean polar bears like to be.

If uninterrupted, a bear may feed for an hour or more. If the kill is small or the bear is large and hungry, he may eat most of it, but often there is a fair amount of fat and meat remaining (p. 127). Unlike brown bears, polar bears do not cache the remains of their kills to feed upon later. Occasionally they will make a few scratching motions in the snow, reminiscent of a dog burying a bone, but most simply depart, leaving the remains on the ice. These carcasses are then scavenged by other bears, particularly subadults, which rely on this source of food as they are learning to hunt for themselves.

The other well-known scavenger of the sea ice is the arctic fox. In the winter, fox tracks may be seen following polar bears for many kilometers far out on the sea ice (pp. 72–75). The remains of almost every old kill are scavenged by arctic foxes and sometimes one will see a dozen or more of them at a time. The foxes are also not shy about trying to help themselves before the bear is finished. As bits and pieces of seal meat begin to be spread about on the ice, the foxes dart in to share the booty. Near the land in places like Radstock Bay, foxes go out on the ice to scavenge during the summer as well. Then they must compete with birds like glaucous gulls, which, like vultures, land within minutes after a seal is killed (p. 127). Usually, the bears ignore the foxes, but occasionally one will make a short charge at a particularly audacious animal that is trying to help itself too soon.

If a bear investigating a new kill finds a bear of similar size feeding on it, they probably will not fight. The chances of mutual and serious damage are too great. Instead, they will feed side by side until the carcass becomes severed. Then each will consume its own piece.

An adult female with cubs will usually avoid adult males, possibly because of the threat of predation on the cubs. If approached by an adult male, she will usually desert her kill. However, if a female has gone a long time without killing a seal, she may not be so easily dominated. One adult female with a two-and-a-half-year-old cub twice met adult males double her size while she was feeding on seals she had killed. In the first case, she had already been feeding for 78 minutes and left immediately when the male appeared. She and the cub had probably eaten the best parts of the animal already and the remains were not worth trying to defend. We kept her under continuous observation, and it was five days before she caught another seal. Nine minutes later, a large adult male arrived and briskly walked toward the female and her cub. This time, rather than leave her kill, she charged the male. He lowered his head and charged to meet her. When they separated a few moments later, she was bleeding freely from a wound on her right shoulder and the male was bleeding from his rib cage. The female then charged him again but he did not move. She stopped before reaching him, whereupon he made a similar bluff charge. They

did not come into contact. Then after standing looking at each other for about 30 seconds, the male, the female, and her cub all began to feed on the carcass together. After 21 minutes of feeding, the female and cub left to wash in a nearby pool. When they returned to the carcass, the male chased them away. Meanwhile, a subadult arrived and circled cautiously, waiting for an opportunity to scavenge. After a few more minutes, the male stopped feeding to go and wash. The cub then approached the carcass. The male turned and ran toward the cub, at which point the female charged him again. He retreated and all three bears fed on the carcass again. Finally, the female and cub departed. After a while, they lay down while the cub licked the female's wound. When the male finished feeding, he followed the female and cub but she would not let him get close. After two hours, he went off in another direction.

## Sleeping

During the summer, polar bears spend a quarter of their time sleeping. It is difficult to measure this precisely because sometimes bears fall asleep when lying still-hunting and it is difficult to tell when the change occurs. When sleeping, bears prefer to lie with their sides or backs to the wind. A bear will often dig a pit in the snow on the lee side of a pressure ridge. Blowing snow may cover it completely and it may remain there for several hours or even days (p. 46). During the ice-free period in places like Hudson Bay, bears often sleep in pits dug into sand or gravel ridges along the beach. On the open ice a bear may simply lie on its stomach with its hindquarters to the wind.

One time that a bear always likes to sleep is an hour or two after a meal. In areas near the coast or in places like Radstock Bay in the summer, there are usually hillsides with patches of snow on them. Females with cubs often climb a hundred meters (or yards) up one of these snow slopes and dig a pit big enough for them all to sleep in. Occasionally, subadults may do this as well. From the hillside, they have a good view of the area and are less likely to be surprised by another bear. Adult females with cubs in particular do this, probably to reduce the risk of encounters with adult males that might try to prey upon a cub.

Of seventeen sleeps longer than an hour, the average length was 7 hours and 45 minutes, not much different from what a lot of humans need. There is also a tendency for bears to sleep more during the day than at night, although in the summer period of 24-hour light the difference is only relative. The preference for being active at night may relate to the behavior of the seals. The seals feed at night when their prey species come up closer to the surface of the water. They probably surface more often to feed at night, so a bear's chance of catching one at its breathing hole is greater then than during the day.

## Swimming

Polar bears are renowned as strong swimmers. As they travel, they swim across bays or wide leads without hesitation. In the summer, they may swim for hours at a time for no apparent reason at all. Bears of all ages and sex classes swim, but subadults and cubs do so the most.

I remember watching a large adult male who had a big number on his back. He had been hunting in Radstock Bay for several days when he suddenly changed his direction of travel for no apparent reason. He walked over to the floe edge, jumped into the water, and swam out from shore until he disappeared from sight. Later the next day, we got a radio message from biologist colleagues that they had seen the bear on Somerset Island, about 100 kilometers (60 mi.) to the south across Lancaster Sound.

Polar bears are sometimes seen several hundred kilometers offshore in the Labrador, Greenland, and Barents seas but their presence there is probably accidental. The ice floes drift south in the spring and summer and a few bears are probably carried along with them. When the ice melts, the passengers are left to their own devices. In earlier times, some of these bears swimming in the open ocean were captured by sailors and taken back as gifts to the royalty of Europe.

Polar bears are fairly buoyant because of their body fat. They swim dog-paddle style with their large oarlike forepaws. Their hind legs float out behind, occasionally serving as rudders. They dive in two ways. The most common is to go down headfirst, leaving the rear end out of the water until the thrust of paddling by the forepaws pulls it under. The second, which was only observed during aquatic stalks, is to surface only with the nose and then slip backwards into the water. I once watched a polar bear at Radstock Bay diving for kelp. After bringing a stalk to the surface, he lay on his back like a sea otter picking out pieces to eat.

When swimming, bears sometimes try to capture sea birds and ducks by diving underneath them and biting them from below. Several people have seen birds caught this way. Bears seem to be more successful at this when the water is rough, probably because the wave action helps to keep the birds from seeing the bear when it surfaces to breathe.

## How Polar Bears Affect Seal Behavior

For many thousands of years, arctic seals have been hunted by land predators. The polar bear has been most important but there have also been foxes, wolves, probably a few grizzly bears, and in more recent time, indigenous man. In contrast, the seals that occupy the same sort of habitat in the Antarctic have never had a predator on

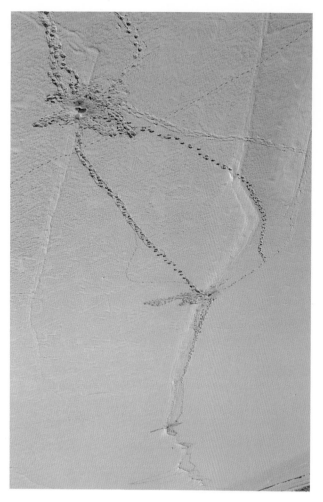

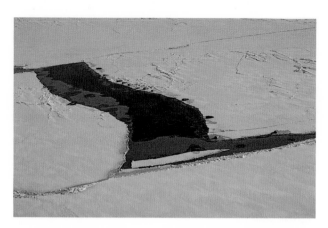

Bearded seals hauled out along the edge of a small lead

An adult male polar bear standing still-hunting, waiting for a seal to surface and breathe by the edge of a lead

Tracks of a polar bear investigating ringed seal breathing holes in young ice. (Photograph © Ian Stirling.)

An adult female doing an aquatic stalk along a water channel melted in the summer sea ice. (Photograph © Ian Stirling.)

*Preceding page:* A subadult investigates a seal breathing hole in rough ice.

Ringed seal breathing hole hidden under a snowdrift, dug out from the side by a hunting polar bear. (Photograph © Ian Stirling.)

A newborn ringed seal pup caught in its birth lair under the snow and killed by a polar bear. The bear probably did not eat this one because it was not fat enough. (Photograph © Ian Stirling.)

Ringed seals lying on the ice constantly raise their heads to look about for polar bears. (Photograph © Ian Stirling.)

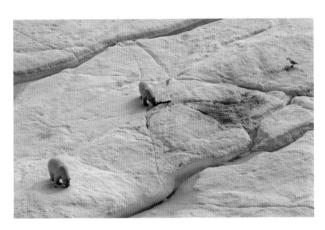

An adult female and her cub feed on a ringed seal while a glaucous gull scavenges. (Photograph © Ian Stirling.)

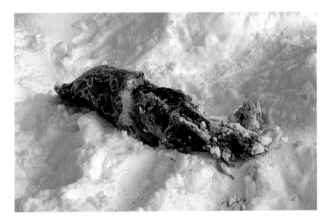

The bear that killed this seal only fed on the preferred parts of the carcass, the skin and blubber. When it was full, it left the remains and went on hunting. The remains of such deserted seal kills are important for young bears that rely on scavenging for survival. (Photograph © Ian Stirling.)

Bearded seal pup only a few hours old. They are born on the open ice with no protection from wind or predators. The white patches suggest this species of seal may be in the process of evolving a white coat for the pups to help them hide from predators. (Photograph © Ian Stirling.)

In these photographs, you can see how the legs move when a polar bear runs.

Polar bears must be agile and have a fine sense of balance for jumping between ice floes. Here a young bear jumps between two floes, then decides he likes the first floe better and returns.

When running at full speed, they gallop but can not maintain this pace for long, especially in the heat of summer.

the surface of the ice. The presence or absence of terrestrial predators has had a profound effect on the behavior of seals, especially those that live on and under the landfast ice. The ringed seal of the Arctic is one of the smallest seals, averaging only about 60 to 70 kilograms (130 to 150 lb.), while adult Weddell seals in the Antarctic regularly exceed 400 kilograms (880 lb.). Both wean their pups in about six weeks, despite the large difference in body size.

Both must maintain their own breathing holes when the ice freezes in the winter. The ringed seal scratches the newly freezing ice with the heavy claws on its foreflippers; the Weddell seal abrades it with its large canine teeth. Both situate their breathing holes in the last cracks to freeze, but there the similarities end. Each ringed seal maintains three to four individual breathing holes. If it only had one, and a polar bear was waiting above it, it would face certain death when it had to surface to breathe. In contrast, several Weddell seals may share the maintenance of the same breathing hole. In an area where feeding is good, Weddell seals may have many breathing holes concentrated in a small area and several seals may be hauled out at a single hole. Ringed seals are widely distributed at low densities and, most commonly, only one seal hauls out at a single hole. If there are more, they may not all be able to escape a charging polar bear. Individual breathing holes are usually more than 200 meters (600 ft.) apart.

When ringed seals bask on the ice near the breathing hole, they appear restless, because they raise their heads every few seconds to look around for possible predators. They retreat to the water at the slightest disturbance. When Weddell seals haul out on the ice, they fall into a deep sleep and move so little that often the outline of the seal's body is melted into the ice. Ringed seals rarely defecate onto the ice, probably because the odor would attract predators, whereas Weddell seals commonly do so. In the open water

season, if no ice is present, Weddell seals haul up on land to rest, since there are no predators. Ringed seals remain in the water until the water freezes again.

The pressure of surface predators on arctic seals has been so great that in most species the young are born with a white coat to camouflage them from their enemies as they lie helpless on the ice. Antarctic seal pups are all born with dark coats. An interesting anomaly is the bearded seal of the Arctic, which gives birth to a pup with dark hair. However, there are white blotches on its head, back, and rear flippers as though it was currently in the process of evolving a white coat for the pup (p. 127).

In the landfast ice where ringed seals are born, polar bears are so abundant that if the pups were simply born onto the surface of the ice they would be extremely vulnerable. The pups are hidden, however, in small birth lairs in snowdrifts over breathing holes. It seems likely that the pressure of predation has kept the ringed seal small because a larger animal would not be able to hide in the snowdrifts.

In 1979, I saw the importance of the subnivean birth lairs vividly demonstrated. I was catching bears in a large bay along the southeastern coast of Baffin Island, an area that occasionally gets heavy rainstorms in early spring. We had a twenty-four-hour downpour and it was another day or so before the weather improved enough to fly. We went back to the same bay and found that the rain had washed away all the snowdrifts. The seal pups were completely exposed when they were only a few days old. The bears simply went from seal to seal, killing them all. From the air, red blotches of blood on the snow were visible everywhere.

Because the ringed seals are so spread out under the ice, a territorial male can probably control access to no more than one or two females. The sex ratio of adult males to females is even. In contrast, one Weddell seal male can exclude his competitors from a single breathing hole when it is being used by several females. In this case, polygyny can develop and the sex ratio in breeding areas is about six to eight females to one male. This, and the other differences already described, illustrate how important the pressure of predation by polar bears has been in molding the biology of the arctic seals.

## Other Hunting

Several people have reported the presence of polar bear scars on belugas (white whales). This has stimulated debate about whether or not polar bears are able to kill them regularly. For most of the year, belugas in open water are not accessible to polar bears. But in about midsummer, large numbers come into the mouths of rivers all over the Arctic.

For several summers, Tom Smith and Becky Sjare studied the behavior of white whales in the estuary at Cunningham Inlet on the north coast of Somerset Island. Large numbers of whales come into shallow channels of the river and occasionally individual whales become stranded on the gravel for a few hours at low tide. On one occasion, Becky watched a large male lie still-hunting at the edge of an ice floe as several whales swam close by. On two other occasions, the bear dove in and managed to kill young belugas about the size of a large bearded seal.

Considering the number and accessibility of belugas at Cunningham Inlet, it is surprising that more bears do not hunt there. A possible reason is that it is difficult for a polar bear to kill a beluga, and only a few large males learn how. Other bears do not often have the opportunity to learn by watching.

At Gascoyne Inlet on the southwest corner of Devon Island, belugas swim very close to the edge of the beach in the summer and occasionally go into the inlet to feed on arctic cod. Buster

Welch, a marine biologist who works in the area, told me of seeing a large male polar bear that appeared to be lying still-hunting on the tip of the point by the entrance to the bay. Although the bear did not attack any whales while Buster watched, he concluded that it was waiting to do so.

During the winter, groups of belugas or narwhals sometimes become entrapped at a breathing hole or in a small area of moving pack when all the surrounding ice freezes. The Inuit call this a *savsatt*. If the weather is cold and the ice continues to freeze around the edges, the breathing hole becomes smaller and crowds the whales into a diminishing area. Sometimes this presents a bear with a hunting opportunity. In 1974, an Inuk hunter from Grise Fiord found a beluga that had been killed by a polar bear at a small *savsatt*. There is also an eyewitness account of a polar bear killing a beluga near Novaya Zemlya in the Soviet Union. The bear struck the whale on the head with a forepaw and then dragged it up onto the ice. It seems possible that the scratches reportedly seen on the backs of belugas came from polar bears hunting at *savsatts*.

The most spectacular account of polar bears hunting belugas at a *savsatt* was reported in the northern Bering Sea, just south of Bering Strait, by the Alaskan marine biologist Lloyd Lowrey and two colleagues. Polar bears had captured a minimum of forty belugas and dragged them out on the ice. Because of drifted snow, a total count of dead whales could not be made, but possibly there were an additional twelve to fifteen carcasses. Photographs showed five bears feeding on the carcasses and observers thought there could have been up to fifteen polar bears in the area. It was apparent from their grey color that most of the dead belugas were young animals. Only a small amount of feeding had taken place on most of the carcasses.

Sometimes polar bears also kill walruses. Most

of the animals killed are young since the adults weigh 1,100 kilograms (2,500 lb.). With their lethal tusks and thick armorlike skin on the neck and shoulders, walruses are formidable opponents. In fact, there are accounts of polar bears being killed by adult male walruses.

In our work around the polynyas of the High Arctic, we have found a number of walruses killed by polar bears. We have also found extensive quantities of blood on the ice where attacks took place but the animals got away. Groups of female walruses with calves are quite alert when hauled out on the ice. At the slightest disturbance, they head for the water. The only animal they could be that nervous about is a polar bear.

Groups of walruses in the water are known to attack hunters in boats, but I once saw a group of walruses give a remarkable coordinated threat to a polar bear. It took place below our camp at the Dundas Polynya. There was one large pool of water a few hundred meters across, and several small ones along the northern edge. Nineteen walruses were hauled out at the smaller holes in groups of one to six, with a few more animals in the water.

They attracted the attention of a small female polar who weighed 154 kilograms (330 lb.). As she slowly approached one group after another, they saw her and went into the water when she was 50 to 75 meters (50 to 75 yds.) away. In the water, walruses began to rise up and peer in the direction the bear was coming from before it was possible for them to see her. As she neared the edge of the largest pool, the water was calm and no animals were in sight. Suddenly, twenty or more walruses surfaced simultaneously several meters from the edge of the ice and charged toward her. The group suddenly stopped in front of her and two large individuals in the front of the group did a rapid forward-rolling dive and smacked their rear flippers on the water. The sounds were loud, like gunshots. The bear immediately turned and ran 100 meters or so to the

north at a gallop. She then slowed to a walk and proceeded to leave the area.

This observation has a couple of intriguing aspects. First, it appeared that the walruses in the water might be using underwater calls to warn animals in adjacent pools about the bear. Unfortunately, we could not be sure because our hydrophone was not in the water at the time. Second, when threatened with attack, they mounted a remarkable and coordinated response.

Considering Alaskan brown bears' well-known taste for salmon, it is surprising that polar bears do not prey much on arctic char, which run in shallow rivers all over the Arctic. I have heard isolated accounts of individual bears that seemed to be fishing but nothing that indicates an established behavior. Since polar bears are so quick to learn about temporary food sources, such as dumps or hunting camps where seal carcasses may be left, it seems likely they would show the same fidelity to a fish run if they had ever used it.

There is one possible exception but we do not have enough information to properly evaluate it. As late as the 1770s, the explorer Cartwright reported "large numbers" of white bears in July and August fishing for salmon in the Eagle River to the west of the present-day settlement of Cartwright on the Labrador coast (see Fig. 1). Apparently he found well-defined trails in the bush along the riverbanks and the remains of partially eaten salmon, an account similar to present-day descriptions of areas where brown bears feed on salmon in Alaska. In July 1778, thirty-two white bears were seen feeding on salmon along with three black bears. A number of bears were shot and more would have been if the hunters had not run out of ammunition. For several years, bears were killed there for their hides, but few records of numbers were kept. In recent years, few bears have been reported as far south in Labrador as Cartwright. Whether or not polar bears had once learned to fish in that area remains a mystery.

Finally, there are stories of polar bears using blocks of ice or stones to kill other animals, particularly walruses. Perhaps most famous is the engraving in C. F. Hall's *Life Among the Esquimaux* that shows a polar bear on top of a cliff hurling a large stone down at an unsuspecting walrus. According to Hall's Inuk companion, this did not kill the walrus but only stunned it long enough to give the bear time to run down and finish the job.

In a variation of this story, the bear carries a block of ice on its shoulder and then stops to hurl it at another animal. Although the stories are widespread, witnesses are hard to come by. Various people, including the polar explorer Frederick Cook, tell of seeing bears stand up holding objects. Indeed, anyone who watches polar bears performing in the Moscow circus will see the same thing. An observation recorded by the arctic physiologist Kare Rodahl offers a possible clue to the mystery. He saw a polar bear that narrowly missed a bearded seal and then "leaped up onto the floes and in his fury began to toss lumps of ice about." I have also seen bears show frustration after not getting a seal. Sometimes they will swat the snow, or a female will swat a cub that disturbed her hunt. I saw one bear stalk a ten-gallon oil drum on the ice for half an hour. When it saw what it had caught, it gave it a cuff that sent it spinning several meters across the ice.

Polar bears are intelligent animals, and thus I would be reluctant to say one could not learn to use a "tool" in this manner. However, if they do, it is a rare occurrence. My guess is that blocks of ice found around hunting sites were broken off in a moment of high frustration.

# Life and Death

As Charles Darwin pointed out in *The Origin of Species,* intense competition for survival is a daily fact of life. This is as true for polar bears today as it was when it was first written over 125 years ago, regardless of how tranquil the bears may appear sometimes.

From the time that each year's crop of cubs leaves the den to begin life in the wind and snow of a cold arctic spring, weaker individuals continue to fall by the wayside. Actual observations of polar bear deaths, in which one can be certain of the cause of death, are rare. However, like detectives, we can gain considerable insight by carefully sifting through the wealth of anecdotal and scientific information that has accumulated.

## Mortality of Cubs

For cubs, the struggle for existence begins immediately. Some cubs do not live long enough to leave the dens they were born in. A few are born dead or die shortly after birth. Mitch Taylor, Ray Schweinsburg, and Thor Larsen summarized instances of infanticide and cannibalism in polar bears from all over the Arctic. They found seven reliably documented cases in which the mother bear was so malnourished that she killed one or

both of her own cubs and ate them in order to survive herself and breed again. If there are seven documented instances of an event as difficult to observe as this, it may occur reasonably frequently. In terms of survival of the species, the best strategy for the mother bear is to maximize her own survival by eating the cubs and waiting until next year when she may be in better condition and have a better chance of raising strong, healthy cubs.

A few cubs, although strong enough to leave the den with their mothers, are still too small and weak to compete with their siblings in the harsher environment outside. The mortality rate of cubs in their first year of life varies between 20 and 40 percent in different populations.

Occasionally we see the results of environmental pressures at work. When females leave their maternity dens with their cubs to hunt seals on the sea ice, they have usually exhausted their stored fat reserves. Catching a seal soon after reaching the ice is probably critical to her survival, as well as that of her cubs. In most areas, the maternity dens are within a few kilometers of the landfast ice. Thus, families expend little energy returning to their hunting habitat. The date that polar bear families leave their dens for the sea ice has probably evolved to coincide with the period when the ringed seal females are having their pups in birth lairs. This increases the chance of hunting successfully and improves the survival of the cubs.

Polar bears are creatures of habit and females with cubs return to feed in areas where they have hunted before. However, shifts in the distribution of the seals may take place between years, or the numbers of pups born may be greatly reduced.

For example, in the spring of 1974 the production of pups in the eastern Beaufort Sea fell so sharply that some polar bears were threatened with starvation. We caught two emaciated adult females soon after they had left their dens. They had been unable to catch any seals and had lost their cubs. Another female had cubs with her that were so thin that one could barely walk. The following year, ringed seal pup production fell even lower and did not increase again for a year or two after that. As a result, bears of most age and sex classes were significantly lighter in 1974–75 than they were in 1971–73.

For the next ten years, as we observed the age structure of that particular polar bear population, we noted that animals born in those years continued to be poorly represented. Clearly, the cubs born from about 1974 to 1976 did not survive well. Some of the adult females were in such poor condition that they failed to have cubs at all.

In general, females with cubs tend to seek out habitat different from that favored by adult males; they avoid contact with males when their paths do cross. This is probably because adult males sometimes kill cubs and eat them. Undoubtedly this happens, but after several thousand hours of watching polar bears at Radstock Bay over a number of years, I am not convinced it is a frequent occurrence. We have seen many adult males follow family groups, sometimes for hours at a time, without being able to get any closer than the female was willing to tolerate. On the Hudson Bay coast in the fall, females with cubs generally avoid males, but we have also

seen several instances where they attack and dominate the males.

Large adult males are well insulated, slow moving, and overheat quickly when they run. Young bears and females can probably run for much longer without overheating. On the other hand, adult males often take killed seals away from smaller bears, including females with cubs. Although we have occasionally seen females defend a kill against an adult male, this involves the risk of serious injury and is usually avoided. Occasionally, an adult male might surprise a family that is sleeping or feeding. In this circumstance, the male might be able to make a rush and capture a cub. In such cases, the female might try to defend the cub, risking her own death in the process. In 1984, two separate incidents were reported in the Canadian Arctic in which an adult male was found feeding on the fresh carcass of an adult female while her yearling cubs watched from a distance (p. 152). Biologists thought that each female had died in defense of her cubs.

I have stressed that polar bears are intelligent carnivores that learn things quickly and that each animal is an individual with its own accumulated set of experiences. Of course, this applies to other carnivores as well. In a few instances, wolves have learned to prey on bear cubs of various species. A few times bears have also taken a kill away from some wolves and a defender has been killed or injured in the process. In 1983, Malcolm Ramsay and I found evidence of a pack of wolves, in the large denning area south of Churchill, that had learned to kill polar bear cubs when they were on their way to the sea ice from their maternity dens.

Tracks in the snow showed that some wolves would attack and distract the mother while another would seize a cub and escape with it. The cubs were devoured completely. 1983 was the first year in which we found evidence of this behavior, and I have not heard about it from anywhere else. In most areas, the distributions of polar bears and wolves do not overlap very much. We did not see any evidence of wolves continuing to hunt polar bears in this area again until 1988.

## Mortality of Subadults

The most difficult period of a polar bear's life falls between when it is weaned at about two and a half years of age and the time it becomes a successful adult at about five or six years. In this period, they are subadults, the teenagers of the polar bear world so to speak, and they have the polar bear equivalent of all the insecurities associated with human adolescence. They are fairly inexperienced at hunting and do not kill seals as often as adults. Worse, when subadults make a kill, they have a greater chance of losing it to a bigger bear. I have seen a number of subadults forced off their own fresh kills and reduced to scavenging on the scraps left by the dominant bear.

Although we have caught a lot of healthy subadults over the years, we have not caught many fat ones. This is the opposite of the pattern in adults. Overall, younger animals go into the winter in poorer condition than adults; thus their greatest mortality factor is probably starvation.

This is why so many of the so-called problem bears that haunt human camps and garbage dumps are subadults. They are thin and hungry. Trying to find something to eat around a human camp is preferable, at least in the short run, to starvation. In the process, they may threaten someone's life and end up being shot. It is unusual to find dead bears in nature, although I know of two subadults that died naturally. They were found tucked in behind pressure ridges in the spring, curled up in a ball and frozen solid the way they died. Autopsies showed them to be little more than skin and bone with no body fat, losers in the continuous struggle for survival.

## Longevity of Adults

Once polar bears reach maturity, they seem virtually immortal. Annual survivorship of adults could be as high as 95 percent in some areas. They have no natural enemies except, in rare circumstances, each other. They have learned the most efficient annual cycle of hunting, movements, and fasting in their home range. The importance of this learning cannot be overemphasized, especially when one considers how different environmental conditions are between areas such as southern Hudson Bay, the polar basin, and East Greenland.

It has also been necessary to learn how to avoid human hunters. For example, in the spring of 1987, I caught an adult female in a heavily hunted area, within a few kilometers of where I had first tagged her sixteen years earlier. She had hunted in the same area in the spring every year and learned how to avoid capture.

No matter how canny a bear may become, survival is still a constant struggle. The battles between adult males for reproductive rights produce scarred heads (p. 41) and broken teeth, providing strong testimony to the intensity of the competition. This is probably the main reason that adult males have a higher annual mortality rate than females. The average age of adult bears in a healthy population may be around nine or ten years. In general, only a small proportion live past about fifteen to eighteen years. Most adult females cease having cubs after the age of twenty, although a few manage it.

The oldest polar bears we have aged from the wild were both thirty-two-year-old females. One was from the islands of James Bay at the very southernmost part of their range. We have little information on those bears, but there seem to be quite a few old-timers. This may be because they are seldom hunted there. The oldest male we have aged from a wild population, twenty-eight years, was also from James Bay.

I will never forget the other thirty-two-year-old female. Tom Smith, his Labrador retriever Bug, and I were continuing our seemingly incessant search for seal breathing holes and birth lairs in different habitats. This time, we were at Radstock Bay, on Devon Island, where we have done much of the behavioral observation on polar bears. It was mid-April and cold but clear with a light wind—beautiful for that time of year. I was climbing through some large pressure ridges close to the shoreline when I saw a polar bear lying in the snow about 2 meters away. You don't stop to think at such moments and in an instant I was well out on the sea ice. At that point, I had time to think. Something in my subconscious told me that the bear's behavior in the pressure ridges was unusual. Also, it was strange that it had not stood up to peer out at us. Carefully, and with Tom covering me, I crept back into the ice blocks and drifted snow until I was only a few meters from the bear again.

She appeared small and at first I thought she was a subadult. Then she lifted her head and slowly looked in my direction, with eyes that no longer had the sparkle and alert intensity of a normal polar bear. She had a medium-sized frame but appeared small because her body was almost wasted away. Every bone pushed up under her hide, making it appear as if it had been draped over a skeleton. The hair over the worn muzzle was missing in a few places. Only then did I realize the bear was not a subadult. She was a very old adult female in the last stages of starving to death (dying of old age some might say). She could neither attack me nor try to escape. She couldn't even stand up.

Although I have seen thousands of polar bears and studied them for years, that moment was especially poignant. Here, in an arctic snowdrift, this ancient matriarch was ending a long, experience-rich life. Much of it had probably been spent within 160 kilometers (100 mi.) of where she now lay. I had probably watched her hunt seals

and nurse her cubs on sunny summer days in better times. As she rested her head on the snow, I watched her quietly and thought about the life she must have led. Then I left her to die undisturbed.

There was a blizzard the next day, and after that we could not find her carcass in the new snowdrifts. The following summer, we relocated her remains on the beach and collected her skull. When we determined her age from one of her teeth, we found she was thirty-two years old.

Although finding old polar bears in their last months of life is uncommon in most areas, we see one or two on the western coast of Hudson Bay each autumn. They are usually ancient males (p. 176). They are fairly inactive, mind their own business, and are largely ignored by other bears. Occasionally these walking skeletons die on land and we find their remains curled up in a clump of willows or behind an esker along the beach. From the teeth we have aged, it appears that most of them are in their early to mid-twenties. But most survive long enough to make it back onto the ice. The fact we find so few old bears dead along the shoreline suggests to me that most die on the sea ice in the winter.

In captivity, polar bears enjoy a much longer life than they do in the wild. The oldest male in a zoo that I know of was a forty-one year old in London, twelve years older than the maximum recorded from the wild. A male in the Milwaukee zoo reached thirty-four. At least fourteen polar bears are known to have lived more than thirty years. A female in the Detroit Zoological Park turned thirty-eight recently and is still healthy. Before her, the oldest female in captivity was thirty-four.

Considering that few wild female polar bears reproduce after their early twenties, some of the ages at which captive females have had cubs are quite impressive. In 1935 a bear at the Milwaukee zoo gave birth to her last cub at twenty-four years. Katherine Latinen from the Detroit

Zoological Park reported that females aged thirty-four, thirty-six, and thirty-seven all gave birth to live cubs. The thirty-six-year-old female successfully reared her cub while that of the thirty-seven-year-old female died after about a day. As the quality of care, and especially nutrition, of captive animals continues to improve, we will learn more about their genetic potential for longevity and reproduction. The difference between the longevity of captive bears and their life span in the wild gives us some insight into the cost of surviving in nature.

## The Influence of Hunting on Survival

How long polar bears survive is influenced by where they live. Although differences in the environment are important, death by gunshot is probably the greatest single cause of mortality in wild polar bears. The Inuit have hunted polar bears for centuries and their traditional right to continue doing so, within safe, scientifically established limits, is protected by the International Agreement on the Conservation of Polar Bears.

Establishing those safe harvesting limits is more easily said than done. For example, in the Beaufort Sea, before the establishment of quotas in Canada in 1968 and the cessation of sport hunting in Alaska in 1972, there were few restrictions on the numbers of polar bears shot. The unstated assumption was that the population could support the harvest. However, by the early 1970s, the average age of the polar bears being harvested in Alaska was only about five or six years. When I started work in the eastern Beaufort Sea, only 7 (4.1 percent) of the 169 bears I caught between 1971 and 1973 were fifteen years of age or older. A few years later, between 1977 and 1979, after the restrictions had been put on hunting, 15 (10.3 percent) of 145 bears were fifteen or older: twice as many. By 1985–86, 61 (15.0 percent) of 406 bears captured were in that

age group. This improvement in survival has been brought about by drastically reducing the size of the harvest.

So far so good. But this is where it becomes difficult. As discussed earlier, estimating the size of a subpopulation of polar bears is not easy. Usually, we work with what we feel are reasonable yet conservative estimates. In this way, the gamble is that if we make a mistake, it will be on the bears' side. It is not very difficult to calculate the reproductive rate of the female bears. Estimating the annual survival rate, especially of cubs and subadults, is more difficult. We can calculate a survival rate from the age structure of a population. However, to do this, we have to assume that the population and the age structure are stable. In nature, this is rarely true. If the sample sizes are large enough, and have been collected over a sufficient number of years, then mortality rates can be calculated on the basis of capture and recapture records of known individuals. Unfortunately, there are probably only two populations in which this approach could even be tried.

Formerly, we thought that a polar or grizzly bear population could be harvested safely at a maximum level of about 5 percent if the harvest was spread evenly across all the age and sex classes, excluding females with cubs of the year. With advances in computer modeling, we can make detailed estimates of population size, the age structure, reproductive rates, and survival projections. We can alter values to see how sensitive bear populations might be to differences in harvest practices. A large number of tests can be done in a very short period of time. It can be mentally exhausting because of the speed with which the computer spits out new information to think about.

Fred Bunnell and David Tait at the University of British Columbia in Vancouver were pioneers in this approach to wildlife management. Together we did some of the first population model-

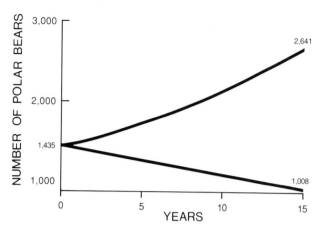

Fig. 6. Population modeling diagram

ing with polar bears. Figure 6 gives an example of the tests one can do.

We started with a population of 1,435 polar bears and entered the estimates of survival and reproduction. Then we subtracted a harvest of 72 bears, or 5 percent of the population. In the first simulation (bottom line), hunting was allowed in both the fall and spring. The quota was removed proportionately from the total population one year of age or older. In the second simulation (top line), we modeled a hunt in which bears could be taken only in the spring. We also provided 90 percent protection for females that were five years of age and older, and cubs from zero to two years. This second simulation protected pregnant

females (which hunters cannot distinguish from young adult males) during the time they are looking for maternity dens in the fall. Adult females with cubs of any age were protected in the spring. Reproductive failure or loss of cubs was set at 10 percent and resulting lone females were allowed to be hunted.

The differences between the two simulations were dramatic. In the first, where bears of both sexes over one year of age were harvested, the population declined to 1,008 in only fifteen years. In the second simulation, which protected pregnant females and family groups, the population increased to 2,641 in the same period of time.

Polar bear females take several years to reach maturity and then they produce cubs at a fairly slow rate. This simple illustration of how sensitive a population can be to the harvest of adult females and family groups shows why hunters should be encouraged to harvest males instead. Recent computer projections by Mitch Taylor, the polar bear biologist for the Northwest Territories, suggest that no more that 1.5 to 2 percent of the adult females in a population can be harvested before it may begin to decline. These results show why it is important to continue to monitor and model harvested polar bear populations closely.

# What Makes a Polar Bear Tick?

The Arctic, like the desert, has a highly variable and often severe climate. For much of the time, the life of a polar bear is one of uncertainty about where the next meal is coming from. In the late spring and early summer, the temperature is relatively mild and seals are abundant, but still, polar bears do not have time for relaxation. They must use every day when hunting is good to prepare for the times when stored body fat is the key to survival. This chapter is about some of the remarkable ways in which the physiology of the polar bear has adapted to cope with the arctic environment.

Take a moment to create a mental picture of the polar bear's environment. During the winter, subzero temperatures prevail for months at a time, made more severe by incessant wind. Periodic blinding snowstorms hinder vision. The sea ice is a highly variable environment, alternating between being frozen during cold calm weather or broken up with large areas of open water during polar gales. In some parts of the Arctic, the ice may drift several kilometers a day. Annual patterns of freeze-up and breakup can vary, causing changes in the distribution and abundance of seals. The result is a continuing search for prey in a changing environment.

To survive in this environment, the polar bear has two major physiological problems to solve: keeping its body at the right temperature and storing enough energy to last between meals that could be a few days or a few months apart. Norwegian scientist Nils Øritsland and several of his students built a unique physiology lab in a deserted military building near Churchill where they collected most of what we know about how a polar bear regulates its body temperature. Nils is a resourceful scientist who seems to think like a polar bear and has a sixth sense of what makes them tick inside. One of his ingenious creations was an enclosed respiration chamber big enough to hold a large polar bear. It had instruments for measuring the oxygen and carbon dioxide in the air breathed by a resting or exercising bear. They put a treadmill in the chamber and monitored the temperature, heart rate, and breathing of bears walking at different speeds. From their discoveries about the bears' metabolism and energy requirements, we gained a greater understanding of their behavior and ecology.

## Body Temperature

The normal body temperature of a resting polar bear is about 37° C (98.6° F) which is normal for a mammal. A bear's fur, tough hide, and blubber layer, which can be up to 11 centimeters (4.5 in.)

143

thick, provide such excellent insulation against the cold that the bear does not have to change its metabolic rate (in other words, burn any more energy) very often. Its body temperature and metabolic rate stay at the normal level even when the thermometer drops to $-37°$ C $(-34°$ F). As long as a bear is relatively inactive, and is not exposed to wind, it does not burn excessive energy in cold weather.

The negative aspect of being so well insulated is that the bear overheats quickly. Just think of putting on all your warmest winter clothes and climbing a steep hill on a hot sunny day and you will appreciate his problem.

## Temperature Regulation

At temperatures ranging from about $-15°$ C to $-25°$ C (about $-4$ to $-12°$ F), a polar bear's body temperature remains fairly constant at walking speeds of up to about 4 kilometers per hour (about 2.5 MPH). After that, however, it begins to climb rapidly until by about 7 kilometers per hour (4.2 MPH), it is almost $39°$ C $(100°$ F), which is a fever temperature in a human. To move at this modest speed, a bear burns up thirteen times as much energy as it would if it was lying down. From these results, it is clear why polar bears plod across the sea ice so slowly as they search for seals. They have adjusted their rate of movement so that energy is used efficiently and overheating is avoided.

One of the bears that Øritsland worked with on the treadmill frequently moved off for short periods of time when the speed was increased, and he sometimes lay down and refused to walk at all! He also growled if the treadmill was run at too slow a speed for him to develop a comfortable pace.

Jimmy Memorana, an Inuk hunter from Holman, once told me that in the old days, if a hunter found the fresh track of a large male bear when hunting with his dog team, he could easily run him down on foot in a few hours. At the time I was a bit puzzled, but now it is quite clear. Large male bears use so much energy to move, and they are so well insulated, that they quickly overheat when they try to run. They are soon reduced to a walk, even in very cold weather.

Polar bears regulate their temperature by both physiological and behavioral means. Measurements made with infrared sensors by Øritsland on an exercising polar bear revealed "hot spots" on the muzzle, nose, ears, footpads, and insides of the thighs. These are the points where the bear dissipates excess heat. Dissections of dead bears revealed that they also have thin layers of muscles, which are richly supplied with blood vessels, lying only a few millimeters under the skin, in the shoulder region. This network brings warm blood from deep in the body to the surface where some of the heat may be given off.

More recently, an American physiologist, Ralph Nelson, used infrared equipment to examine the shoulder areas of resting animals in air temperatures a little below freezing. He found that under those conditions, the shoulders did not give off enough heat to be detectable. That reminded me of a photograph I saw a few years ago that was taken by an Alaskan scientist. He was trying to find out if infrared photography could be used to detect polar bears out on the ice. To test the idea, he found a bear and took some pictures. The bear was so well insulated that it gave off no detectable heat at all. But there was a spot on the infrared photo, just ahead of the bear's head . . . made by its breath! Not every creative idea achieves a breakthrough.

The conductivity of water is about twenty times that of air, so swimming is a good way to cool off quickly. On a hot day in summer, we often see bears swimming along the Hudson Bay coast. A bear running away from a helicopter will often dive into a lead or a lake when it becomes overheated.

On the other hand, it is possible that cooling from extended swimming could be dangerous for small cubs that are only a few months old. Øritsland found that forced exercise did not elevate the cub's deep body temperature for an hour after it had been swimming. This probably explains why females with young cubs are reluctant to swim across leads with them. The cubs would become too cold. Females with cubs older than six months do not show the same hesitation.

When sleeping or lying, bears adopt different postures depending on whether they want to get rid of heat or conserve it. On warm days, they will sprawl out and sometimes lie on their backs with their feet in the air (p. 47). On colder days they will curl up, sometimes covering the heat-radiant muzzle area with a paw. Sleeping on a warm day in the shelter of a pressure ridge, a bear may sprawl over and around the irregular ice blocks (p. 51), looking more like a jellyfish than the ultimate arctic carnivore.

On the western coast of Hudson Bay in the summer, bears spend most of their time lying around doing nothing. There is no point wasting energy when there are no seals to catch. If a bear gets hot, it costs energy to pant, increase the heartbeat, and metabolize water to cool off again. Consequently, they remain fairly inactive, especially on warmer days.

When obese pregnant females first come ashore, they usually do not begin to move inland to the cool earth dens of the maternity denning area until the sea breezes drop the air temperature, or until it begins to rain. Clearly, they aim to avoid the high energy cost of moving a heavy, well-insulated body on a warm day.

## The Cost of Walking

Another fascinating result from the treadmill studies done by Øritsland and one of his students, Ricki Hurst, relates to how much energy it costs for a polar bear to move. There is a general formula one can use to calculate how many calories a four-legged mammal must expend in order to run at a particular speed. Most species fall close to the predicted value for their weight. However, there are some interesting exceptions to the rule. The polar bear is one of them; it uses more than twice as much energy to move at a particular speed than do most other mammals. The low efficiency of movement of the polar bear may be a result of its bulky build and massive limbs and paws. They contribute a sideways motion to the bear's forward movement. This helps to explain the polar bear's preference for lying and still-hunting. It is energy efficient in an environment where calories can be hard to come by. It also explains why bears rarely hunt musk-oxen, even though their ranges overlap in several areas. The energy cost of chasing musk-oxen is probably too great. It is interesting to note that another mammal that requires about twice as much energy to move as predicted is the lion. It too lies in wait or stalks slowly before making only a short charge after which it quickly gives up if it is unsuccessful. These brief examples illustrate how much an understanding of the physiology of these two predators helps us to interpret their behavior and ecology.

Knowing how much energy it costs a polar bear to run, and how quickly a bear can heat up, also helps to solve another puzzle. There are several large colonies of snow geese along the Hudson Bay coast and during the summer when they moult, these birds are all flightless. Although polar bears have often been seen walking through some of these goose colonies, they almost never chase them. Using some of the results from the treadmill studies, one of my students, Nick Lunn, worked out the likely explanation. To gain the calories contained in an average snow goose, a 320-kilogram (700-lb.) polar bear must catch it in only 12 seconds of running. Otherwise, it will cost him more energy than the bird would yield.

Flightless geese can still run fast and swim. Although devious, they would certainly be caught in the end. Yet even young bears know better than to spend energy chasing them.

## Heart Rate

Øritsland and another of his students, the late Robin Best, examined heart rate in several polar bears. Unlike humans, they found that the heartbeat of resting bears was quite irregular. One unweighed female polar bear had a heart rate that varied between 53 and 64 beats per minute when she was sleeping undisturbed at an air temperature of −28° C (−20° F) on a cloudy day. Another sleeping female had a heart rate that varied about an average of 80 beats per minute on a sunny day when the temperature was just above freezing. Four 230-kilogram (500-lb.) males at unspecified air temperatures had a sleeping heart rate of 33 beats per minute. When awake but lying down, the heart rate in these same bears averaged 46 beats per minute and it increased to 58 beats per minute when the animals sat up. At walking speeds that varied between about 3 and 6 kilometers per hour (1.5 to 3.5 MPH), the heart beat increased more rapidly in relation to speed, rising to about 148 beats per minute.

## Digestibility of Ringed Seals

It may be several days between meals for a polar bear, but when it does eat, it makes up for lost time. When a bear kills a seal, it may eat for an hour or more with only short pauses to wash or look around. Captive bears in zoos can easily eat 10 percent of their body weight within 30 minutes. There are records of polar bears killed in the Soviet Union with 10 to 71 kilograms (22 to 156 lb.) of food in their stomachs, although the size of the bears was not recorded. Robin Best esti-

mated that a bear's stomach capacity could hold 15 to 20 percent of its body weight.

Best determined that polar bears, like other carnivores, had high digestive efficiencies for the principal dietary components of protein and fat. The polar bear assimilates an impressive 84 percent of the protein it eats and 97 percent of the fat, for an overall energy intake of 92 percent of that which is available in the diet. According to Best, an average active adult polar bear would need about 2 kilograms (11 lb.) of seal fat per day to survive. This means that any seal more than a month old could satisfy a polar bear's needs for a day and an adult seal could provide enough energy for about eleven days, while the bear continued to hunt.

This explains why the bears hunt the fat, newly weaned young seals so intensely in the late spring and early summer. It is the bear's annual opportunity to lay down the majority of the fat stores it will need to survive the open-water period of late summer. The pregnant female polar bears of southwestern Hudson Bay must put on enough fat to survive not only the open water period of late summer and autumn, but the whole of the maternity denning period as well, a total of about eight months. Some of these females come ashore carrying well over 200 kilograms (440 lb.) of fat to survive upon.

## Hibernation and Energy Conservation

Hibernation is an adaptation that allows some species of mammals to store up fat when the feeding is good and then burn it off slowly while in a resting state when feeding is poor. This allows them to survive when food is unavailable, usually, but not always, during winter. True hibernators such as some of the rodents, bats, or insectivores have a marked drop in heart rate, their body temperature may approach 0° C (32° F), and it may take some time to wake them up.

Although most species of bears go into dens during the winter, they have some important differences from the so-called true or deep hibernators. From research done on captive bears in hibernation by Ed Folk of the University of Iowa, we know that the heart rates of hibernating black and grizzly bears slow down to 10 to 12 beats per minute, but their body temperature only declines to about 31 to 35° C (88 to 93° F). He found that the heart rate of a polar bear held in an artificial den decreased to 27 beats per minute after about a month. Paul Watts of the Institute of Arctic Ecophysiology in Churchill found that the deep body temperature of two female polar bears hibernating in natural dens during the winter ranged from about 35 to 37° C (95 to 98.6° F).

Bears maintain a much higher body temperature than the deep hibernators because they still need to be able to maintain the physiological demands of pregnancy, birth, and nursing the young. Although the bears sleep soundly, they are easily and quickly aroused and can, if necessary, defend themselves.

The pattern of hibernation and seasonal food shortage in polar bears is quite different from that of black and brown bears. In the first place, only pregnant female polar bears hibernate; the rest of the population is active throughout the year, including the winter. Second, for most polar bear populations, the season of greatest food shortage is the open-water period of late summer and early fall, just when black and grizzly bears are eating most intensively in order to lay on fat for the coming winter.

One of the most interesting chapters in our understanding of hibernation in polar bears originates with Ralph Nelson, now the Research Director of the Carle Medical Foundation at the University of Illinois. As an astute clinical nutritionist, he wondered how black bears could hibernate through the winter at near-normal body temperatures without eating, drinking, or producing any urine or feces. When hibernating, the black bear produces all the water it needs by chemical pathways from fat; then it recycles the products without producing waste materials. He thought if he could figure out how this was done, there might be enormous benefits to humans with kidney problems.

Nelson found that he could define the animal's physiological state by the amount of two chemicals in the blood, urea and creatine. Creatine is produced by normal muscle activity and its level in the blood remains pretty much the same all the time. The amount of urea in the blood, however, goes up when an animal is eating and becomes very low when it stops eating and lives only on its fat.

After some experimentation, Nelson defined the urea-to-creatine ratio (that is, the number of units of urea in a sample of blood divided by the number of units of creatine) of a hibernating black bear as anything less than 10. He then examined blood samples taken from nonfeeding polar bears on the western coast of Hudson Bay during the ice-free period in the late summer and fall. He found that the ratio of urea to creatine was very low in them as well. He suggested that even though they were not in dens, they were (in the physiological sense) hibernating. This gave rise to the rather intriguing term "walking hibernation."

At this point, Malcolm Ramsay thought to combine Nelson's physiological expertise with ours in ecology. He thought this would help Nelson with his medical research, and us to understand more about how polar bears have evolved to live in the arctic environment. Consequently, in the summer of 1984, Ralph and two of his medical colleagues joined our field camp on the tundra. For several days, we brought in polar bears of various sizes for nonharmful experiments while other curious bears wandered by and gazed curiously at the goings on.

As the studies continued through the next couple of years, we found that female polar bears coming out of the denning area in the spring were in a physiological state similar to that of the hibernating black bears. More interesting, though, were the results from the polar bears spending the late summer and fall along the Hudson Bay coast. The majority of them were not feeding, and their blood constituents confirmed that they were in the same physiological state as hibernating black bears. However, at the same time, polar bears that were feeding in the dump had the same blood values as nonhibernating (feeding) black bears. In other words, members of the same polar bear population could be in completely opposite physiological conditions at the same time and place. That is something a black bear cannot do. If you stop feeding a black bear in summer, it will die. If you do the same thing to a polar bear, it appears that it just switches over to hibernating mode and waits until food becomes more abundant again. This seems to indicate that the polar bear has advanced to the point where it can turn the mechanism on or off at will in relation to availability or scarcity of food.

If the biochemical mechanism could be isolated and applied to humans, Nelson feels it could be as important to kidney disease or obesity as the discovery of insulin was to diabetes.

## Physiological Effects of Oil on Polar Bears

During the late 1970s, when offshore exploration for oil in ice-covered waters began to increase rapidly, several of us expressed concern about the possible effects of oil spills on polar bears. Research on fur seals and sea otters, which also rely on their fur for a significant portion of their insulation, had shown that contact with oil could cause death from hypothermia. Although the polar bear receives a significant amount of its insulation from its fat layer, it also depends upon its fur for keeping warm.

After considerable pressure from the Canadian Polar Bear Technical Committee and the IUCN Polar Bear Specialists Group, some funds were raised from government and industry to conduct a small number of tests on four polar bears in the lab at Churchill. Tests on polar bear hides purchased from Inuit hunters showed that oil significantly reduced its insulative value.

Three live polar bears were exposed to oil in a manner that simulated natural conditions in the wild, in that they entered a pool with oil on the surface. The bears showed no aversion to the oil and it stuck to their fur on contact, indicating that if a bear swam across a lead with oil on the surface, it would become fouled.

It was winter when the studies were done. The oiled bears began to shiver and then to lick the unrefined crude oil from their fur in the same manner they lick seal oil from their fur when out on the sea ice. Two of those bears suffered kidney failure and died. The third was treated in the Assiniboine Zoo in Winnipeg for five months. It fully recovered and now lives in a zoo in Japan. No experiments were done on the fourth animal.

The rest, as they say, is history. Few experiments in Canada have been so controversial, or their significance so poorly understood. When news of the deaths became public, the press vilified the scientists for weeks for their perceived inhumane behavior. The large amount of valuable information that was obtained was ignored. While regretting the fate of the two animals that died, I feel their sacrifice was of great importance to the conservation of the species. The response of those bears to having their fur fouled by oil, and the speed of that response, left no doubt in anyone's mind about one thing: oil is bad for polar bears and is capable of killing them. Saving the life of the third bear at the Assiniboine Zoo was equally important because it demonstrated that it is possible to salvage an oiled bear if treatment begins soon enough. In any case, everyone now takes very seriously the danger to polar bears of an oil spill or blowout and plans accordingly.

Normally, adult females with cubs avoid adult males. In this unusual sequence, a female attacks a male, for no apparent reason, and displaces him. First, she approaches and stands looking at him, with her head held low in a threat posture.

Then she charges, but stops just in front of him with her head held low, and gives an open-mouth threat with a hisslike roar as she expels her breath sharply. She returns to her cub and reassures it briefly with nasal contact.

She then stands sideways to the male as her cub peers out from behind. She holds her head low and waves it slowly from side to side in threat before charging again, this time hitting the male briefly on the side of his neck with her jaw.

The male backs up slightly but then holds his ground.

Once more, she threatens with her head low before charging.

This time, she bites his neck and pounds his shoulder with her front paw.

Finally he backs away and the cub becomes confident enough to stand in front of its mother. The male then departs.

An adult stands over the partially consumed carcass of an adult female he has killed. Her yearling cubs cowered a hundred meters or so away. (Photograph © Steve Smith.)

Typical posture taken when two adult males fight.

They face each other, give mutual open-mouth threats, and try to get a good enough grip on the neck to be able to knock the opponent off balance and pin him to the ground. Even when the battle gets rough, there are few injuries in this ritualized behavior.

Occasionally blood is drawn, but such wounds are superficial.

# The Polar Bears of Churchill

The most famous polar bears in the world live on the western coast of Hudson Bay; some of them come into the town of Churchill, Manitoba, every fall. This annual arrival of polar bears in a community has focused the attention of both science and the public on Churchill's bears. They have been the subject of dozens of television programs and hundreds of newspaper and magazine articles. Untold thousands of photographs have been taken by photographers, both amateur and professional, who visit them every year. Over the years, many people have asked me why the polar bears are there, what they are doing, and whether there is anything special about them. In fact, the polar bears of Churchill are quite remarkable and merit a chapter of their own.

Polar bears have been reported in the Churchill area since the first white man arrived there, and they were known to the Indian trappers of the region long before that. The first European to spend the winter in western Hudson Bay was the Danish explorer Jens Munk. He spent the long, cold winter of 1619–20 near the present-day site of Churchill and reported the killing of a polar bear there in September. English traders with the Hudson's Bay Company visited the Churchill area in the summer from 1669 to 1672. In 1682, York Factory was established at the mouth of the Hayes River, about 150 miles southeast of Churchill.

The presence of the maternity denning area north of the Nelson River and south of Churchill seems to have been well known to the Indian trappers of the area. According to the Hudson's Bay records from York Factory, 5 to 100 bear hides were traded there each year, including those of females and cubs taken from the denning area to the north.

In 1930, the Province of Manitoba passed an act protecting the dens of all fur-bearing animals while upholding the rights of Indians to hunt polar bears for their own use. A system of registered traplines was established in the area of York Factory in the early 1950s and trappers were encouraged to limit their kill of polar bears. In 1954, new wildlife regulations were passed that made it illegal for whites to hunt polar bears and for anyone to trade or barter in polar bear hides.

Although the presence of polar bears was well known in the area, they did not attract much attention until sometime after the establishment of a military base at Churchill in 1942. Military land training maneuvers took place on the tundra east of Churchill during summer and fall, when they were most likely to and undoubtedly did encounter polar bears. It is also likely that bears were attracted to the base's garbage. There is no record of how many bears were killed by military personnel in self-defense, for protection of property, or to obtain a hide for a souvenir, but it is reasonable to suppose the number was substantial. In those years, polar bears rarely came into town in the fall as they became famous for doing later.

It is always difficult to determine cause and effect relationships when trying to reconstruct the past. Nonetheless, it seems two events were significant in reducing the number of polar bears killed along the coast and in the denning area between the Churchill and Nelson rivers. In 1957, the Hudson's Bay Company closed York Factory and the resident Indian trappers were moved inland. An average of seventeen polar bears were taken there annually from 1952 to 1958, but since then almost no bears have been killed in the denning area. In 1964, the Canadian Army withdrew from Fort Churchill. After that, the soldiers returned only occasionally for specific training sessions, often in winter when there are few bears on land.

The number of polar bears killed each year probably declined; this likely resulted in an increase in the size of the population along the Manitoba coast of Hudson Bay. It is also possible that the size of the population did not change but because bears were no longer being harassed or killed quite so often near Churchill, they began to occupy the area more regularly. Personally, I favor the first explanation but I admit it is not provable.

Whatever the reason, through the 1960s increasing numbers of polar bears were seen each autumn along the coast east of Churchill. They became regular visitors and by the mid-1960s were a constant sight at the three garbage dumps maintained there at the time. In November 1968, up to forty polar bears were recorded in the dump at one time. People came to view the bears, to feed them, and, regrettably, sometimes to pro-

voke them. People and bears sometimes scavenged side by side in the dump. Bears often wandered into inhabited areas to feed on garbage and occasionally break into buildings where food was stored. Not surprisingly, the number of encounters between people and polar bears increased. People were attacked, but not killed, in 1966 and 1967. On 17 November 1968, a nineteen-year-old Inuk boy who was a student at the school unwisely set off to follow a fresh set of polar bear tracks through an area of stunted spruce trees and undulating rocks near the school. He had separated from his companions to follow a separate set of tracks when he suddenly stumbled onto a sleeping polar bear. The startled animal leaped to its feet, chased the boy, and caught him almost immediately. The boy's friend heard his screams and ran for help. The local Royal Canadian Mounted Police attempted a rescue but could not get a clear shot at the bear without endangering the boy or onlookers crowding around the spot. A shot was fired over the bear, whereupon it ran about 70 feet with the boy in its jaws before dropping him and rearing up on its hind legs. At this point, the bear was shot and the boy was rushed to the hospital, where he died shortly afterward.

There was both concern and confusion about the increasing numbers of bears and consequent encounters between bears and people. While it was clear things had changed drastically, people were less certain why that was so and what should be done in response. Why were the bears coming to Churchill in the first place? Did the same bears keep coming back year after year? Did all the bears in the population come to town or only some of them? If it was only some of the bears, which ones were they? Why were there more bears in some years than in others? How dangerous were the bears? What could be done to reduce the threat to the human population? Could the danger to humans be reduced to an acceptable level without killing all the bears? Did feeding on garbage have a detrimental effect on the polar bears?

These were the circumstances and issues that ushered in a new era of research on and management of polar bears in the Churchill area. In the past twenty years, more studies have been conducted on polar bears in Manitoba than anywhere else in the world. The work has produced fascinating, useful, and sometimes surprising results.

## The Setting

At this point, it is necessary to gain some understanding of the ecological constraints that polar bears on the western coast of Hudson Bay must live under so that we can more fully appreciate their present-day behavior. Southwestern Hudson Bay lies near the southern limit of where polar bears are able to live on a year round basis (see Fig. 7). Conditions there are different from the High Arctic Islands of Canada or the Laptev Sea in the Soviet Union.

The most important factor is that the ice throughout Hudson Bay melts completely by about the end of July or early August and does not refreeze until early November in most years. Consequently, all the bears in the population must come ashore with enough fat stored on their bodies to survive about four months ashore without access to seals until the ice refreezes in early November. For the pregnant females the matter of stored resources is even more critical. At the time when the ice refreezes on the bay, and the rest of the population can go back to hunting seals, they must enter their maternity dens. The cubs are born during the winter and are nursed up to about 10 to 15 kilograms (20 to 35 lb.) before departing for the sea ice again the following March. This means the mother bears have gone for eight months without feeding on seals.

Areas of open water of variable size also form during the summer in other areas of the Arctic,

but not all the sea ice melts. Thus the bears can still move in order to remain with the ice and hunt seals. In those areas where open water does force bears ashore, it is usually for a much shorter period of time than in Hudson Bay.

Another important factor associated with life in such southerly climes as southern Hudson Bay is heat in summer. Polar bears are designed to be comfortable in cold temperatures. When the temperature rises to 20–30° C (75–85° F) during July and August, bears must cope with heat stress. Keeping cool is made very difficult by the fact that the bears must carry a large layer of fat just under the fur to survive on; it seems like the worst of both worlds!

## The Start of Research

At first, of course, very little was known; so the first thing to do was to start tagging individual bears. This work was initiated in 1966 by Chuck Jonkel. At that time, very few polar bears had been drugged by researchers anywhere, so he applied what he had learned during his Ph.D. research with black bears in Montana. Chuck established a trapline of foot snares around the Churchill dump and along the coast to Bird Cove and began to tag polar bears there each autumn. He also used a camp at Knight's Hill to the east of Fort Churchill where he had another half dozen or so snare sites. He trained conservation officers from the Manitoba Department of Natural Resources in the handling and tagging of polar bears. By the time I arrived in 1970, they were helping with much of the work as a routine part of their annual fall bear control program in Churchill.

By tagging polar bears and recapturing them in subsequent years, we hoped to answer the first simple but essential question: Did the same bears come back year after year? We also hoped that Inuk or Indian hunters might return tags from

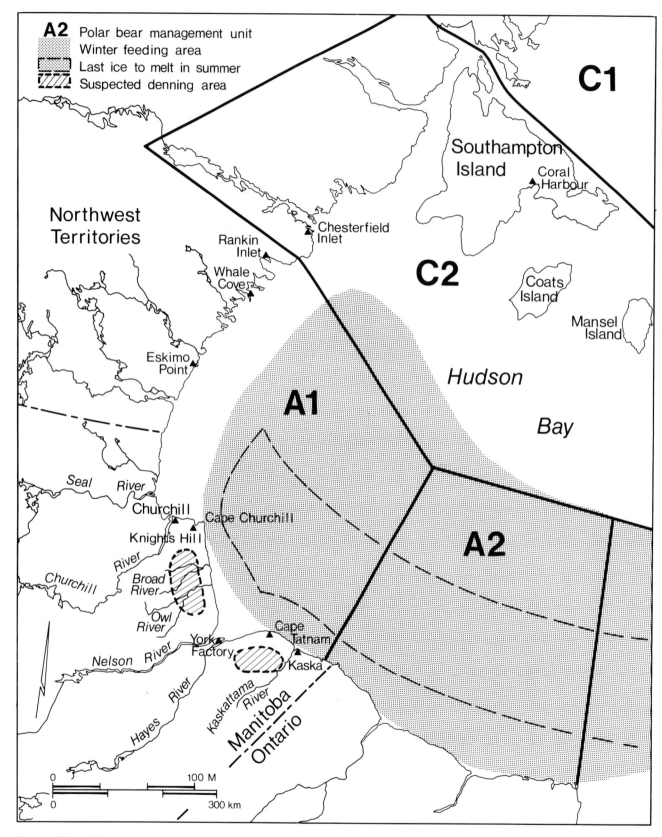

Fig. 7. Churchill and western coast of Hudson Bay

polar bears killed elsewhere, which would give us some idea of the the geographic bounds of the Churchill population.

During the summer and fall from the mid-1960s on, the Manitoba Department of Natural Resources conducted aerial surveys for geese along the coast from Churchill to the Nelson River, and occasionally to the Ontario border. They also counted polar bears, which gave us some idea of the abundance of bears in the area.

From 1967 to 1970, a number of polar bears were fitted with radio collars and then tracked out onto the sea ice after freeze-up. A few collars were also put on pregnant females in dens. When females set out for the sea ice with their cubs, we tracked them from the air. From 1970 through 1976, aerial surveys of the denning area were conducted between February and March, to estimate the productivity of the area.

## The First Ten Years (1966–76)

Right from the beginning it became clear that many of the same bears were coming back to the Churchill dump each year. One bear tagged there as a cub later returned with her own cubs. Several individuals were caught there two or three years in a row.

Some of the bears first caught near Cape Churchill, in the denning area, and south along the coast to the Nelson River, were also recaptured around Churchill. This suggested that all the bears on land in Manitoba during the summer were part of the same subpopulation. Although only twenty-three bears had been tagged along the Ontario coast, and another fifty in James Bay, none was recaptured in Manitoba. Nor had Indian hunters in Ontario or Inuk hunters on the Belcher Islands shot any polar bears that had been tagged in Manitoba. In comparison, Inuk hunters north of Churchill at Eskimo Point, Whale Cove, and Rankin Inlet shot several tagged

bears. Those hunting to the north of them at Chestfield Inlet, or Coral Harbour on Southampton Island, recovered almost none. On the basis of this information, the polar bears from Rankin Inlet in the north to the Ontario border in the south were tentatively considered to be a distinct subpopulation (see Zone A1 on Fig. 7).

Meanwhile, it did not require more than common sense to recognize that Churchill's garbage was the main attraction for polar bears. Meetings were held to find ways to reduce the problem. Public education programs were initiated, dumping of garbage at the town sites was prohibited, and there was discussion about installing an incinerator to burn the garbage. Special cages were built to facilitate the shipping of problem bears by rail or air to zoos in the south to avoid having to kill some of them. There was also discussion about the possibility of establishing a small quota and allowing Indian people to harvest problem bears. The possibility of developing polar bear viewing as a tourist resource was also suggested.

In 1969, the dump near Churchill was closed. All garbage was trucked to the main dump site, which is still being used today, just to the east of the airport at Fort Churchill. Some effort was made to burn the garbage, but this met with variable success, as did a similar plan to truck in sand to bury the rubbish. A plan to move the dump a few miles to the east to the area of Bird Cove was eventually discarded because of cost.

In 1969, Manitoba assigned conservation officers to twenty-four-hour duty to patrol the town and chase out problem bears, at a cost of $20,000. In 1970, the twenty-four-hour polar bear patrol cost $35,000. In 1983, they responded to 191 calls at a cost of well over twice the 1970 figure.

The aerial surveys of the denning area between the Churchill and Nelson rivers, conducted in February and March from 1970 through 1976, turned up some rather spectacular results. Based on sightings of females and cubs, and the tracks of family groups in the soft snow, it was estimated that 100 to 150 cubs were produced in the denning area each spring. Even allowing for the possibility that some bears or tracks might have been counted twice, it was clear it was the largest known denning area in Canada and one of the biggest in the world. It became known as the Owl River denning area since many of the dens were located in the vicinity of the Owl River or to the south of it.

There were a few tracks and dens located in the area of Cape Tatnum to the south as well, but the size of that possible denning area was never determined. No denning areas were reported along the coast to the north of Churchill.

By 1971, it became clear from the tagging studies that bears from anywhere on the coast of Manitoba were part of the same population that was being hunted by Inuit along the coast to the north. At that time, the combined quota for Eskimo Point, Whale Cove, and Rankin Inlet was fourteen bears per year. It was also clear that productivity of the population in the denning area was high.

At this time, Manitoba officials became concerned about the cost of the bear patrols and the possibility of legal action should someone be injured or killed by a polar bear in Churchill. There had already been several court cases in the United States involving attacks by grizzly bears, which generated concern that the same thing might happen in Manitoba with polar bears. Consequently, the Federal-Provincial Polar Bear Technical Committee decided to allow Manitoba a quota of up to fifty polar bears that could be killed or captured to protect human life and property.

This decision set the stage for the possibility of a substantial slaughter of polar bears and raised considerable concern. At this time, the International Fund for Animal Welfare injected itself into the situation. Representatives stated that they wanted to save the lives of problem polar bears that might be be killed in the control program. To do this, they proposed to charter a DC-3 to move the bears away from Churchill. The place chosen as a destination was an abandoned airstrip near the mouth of the Kaskattama River, about 300 kilometers (186 mi.) by air to the southeast of Churchill. There was some discussion about moving the bears as far away as the Labrador coast where it was thought the polar bear population might be depleted. Eventually it was decided to keep them within the normal range of their own subpopulation and avoid possible mixing of different genetic stocks.

Although translocating polar bears was expensive and not thought likely to provide more than temporary relief, it probably did save the lives of the forty tagged animals that were moved out between 1971 and 1975. Certainly, the bearlift gave rise to some hilarious cartoons depicting polar bears in the Churchill airport waiting to catch flights to various places.

The movements of some of these tagged bears after being released also gave some helpful insight into the fidelity of polar bears to the Churchill area. Three males released at the Kaskattama airstrip in 1971 made the journey back to Churchill (480 km or 300 mi. along the coast) in fourteen, fifteen, and twenty-four days respectively. In 1974, it took an adult female with two female cubs of the year just eighteen days to return. In all, thirteen of the forty bears removed were recaptured in the Churchill area, which indicated that the bears knew where they were and that they had a high degree of fidelity to Churchill.

From aerial surveys conducted along the coast during the summer and fall, it became clear there was a general, though not complete, separation of different age and sex classes of bears. Adult males predominated along the coast and especially around the points and small offshore islands. Concentrations could become quite dense at places like Thompson Point, Fox Island, and

Cape Churchill. The fall of 1970 still stands out in my mind as one of the great times for seeing polar bears along the coast. In early November, I saw thirty-six adult males in an area the size of a football field out at Cape Churchill. From a distance, they looked like a small herd of caribou; it took me a moment to even realize what I was looking at. During that same period, we spent four hours flying in an old slow helicopter, on two overcast days with snow on the ground, and saw 239 polar bears!

In general, adult females with cubs or yearlings were less evident along the coast. They preferred the inland areas, often around dens dug into the earth along the edges of lakes or stream banks. In some places, the bears dug pits in the tundra down to the permafrost and rested in them. The frozen ground kept the dens cool on hot summer days and kept insects away from the bears inside. We also found from pleasant experience on a hot, bug-ridden day that going into a bear den provided a welcome escape.

While most of the bears in dens in the inland area were females, it was not their exclusive domain. One warm day in September, Chuck Jonkel and I were measuring the insides of dens and examining them for permafrost. Before entering, we would hover outside the entrance with the helicopter. If a bear was inside, it would peer out at us and we would know not to enter. If no bear was seen, we would still approach with caution on foot but usually find no one home and continue about our work. That particular afternoon, the usual check revealed no occupant, and there did not appear to be much sign of anything when we looked in through the entrance. There was a large pile of fresh dirt and some awesome footprints but nothing else. I climbed into a large chamber just inside the entrance and was glancing about as my eyes adjusted to the dim light when something seemed a little odd. On one side, near the back of the chamber I was in, I could make out something rather white . . . that

moved. The den had an inner chamber and it was occupied! It did not take long to eject myself and get back to the helicopter. This time, after we hovered over the den for a few minutes, a huge adult male lumbered out. When we captured him, we found he weighed 660 kilograms (1,450 lb.) which at that time was the largest male recorded in Canada.

In the late fall, as freeze-up gets closer, more females with cubs and yearlings show up near the coast. Usually they do not remain around the aggregations of large males, although there are always exceptions to any rule. It seems likely that part of the reason the females go inland is to avoid the males on the coast. Male polar bears have been recorded killing cubs, or females trying to defend their cubs (p. 152). Although I suspect this behavior is probably fairly uncommon, it likely contributes to the separation of females with cubs and yearlings from adult males. Sub-adult males and females were found throughout all habitat types and seemed to move about quite a bit.

In the late fall, as the bears concentrated toward the coast in anticipation of freeze-up, they also moved north in the general direction of Cape Churchill. The first ice to form along the western coast of Hudson Bay usually runs north along the coast from Cape Churchill, so it appeared the bears were moving north so as to get onto it as soon as possible. In some years when freeze-up occurred very quickly, it almost seemed as if the bears vanished into thin air overnight. Bears radio tracked out onto the ice for two to three months were still located only 50 to 300 kilometers (31 to 186 mi.) north or northeast of Cape Churchill, indicating that they were remaining in the same general area during the winter.

Meanwhile, as the first research results began to come in, the management problems associated with the bears' annual arrival in Churchill continued. In 1973, an incinerator was built near the dump, but because of various maintenance problems it only functioned sporadically. The public information program continued and so did the patrols by the conservation officers and the Royal Canadian Mounted Police. In 1974, the incinerator only functioned briefly in what proved to be one of the busier years for problem bears. There were 147 bears reported in the residential areas; 11 bears were killed, and 2 were shipped to zoos. In 1975, the incinerator worked better but was only used on weekdays. This meant that garbage had to be stored inside the fence around the facility, and it was not long before the bears learned they could get it. Bears sometimes threatened the driver of the front-end loader while he moved garbage to the incinerator. Slightly fewer bears (119) were reported in the inhabited areas, with only 4 killed and 2 orphaned cubs sent to the Calgary Zoo.

On the basis of the research results, the boundaries of Management Zone A1, from the Ontario border in the south to about Rankin Inlet in the north, were accepted in 1974. Most of the bears along the Manitoba coast returned each year during the ice-free period and spent the winter off the coast of Manitoba and the Keewatin coast of the Northwest Territories. The Inuit in the settlements along the Keewatin coast wanted higher quotas from the population, which it was now clear was shared between Manitoba and the Northwest Territories. Manitoba had a quota of fifty bears for control purposes, although the conservation officers had no intention of killing anywhere near that number of bears if it could possibly be helped. Therefore, in the interest of satisfying some of the needs of the Inuit without unduly raising the overall quota, Manitoba agreed to allow fifteen bears from its quota to be taken in the Northwest Territories. This was an astute move on the part of Manitoba because it provided a temporary solution to a problem without raising the quota on a subpopulation on which there was no reliable population estimate.

## The Second Ten Years (1976–86)

Through the first ten years, we had begun to learn some of the basics about polar bears in the Churchill area. Much of this information was immediately useful in the conservation and management of the bears. It also enabled us to ask some much more fundamental questions about the behavior and ecology of these marvelous animals, questions that would help us better understand polar bears throughout their range. For example, we knew that some of the same bears used the dump in subsequent years. But we also knew there were at least several hundred bears in the population and most of them *did not* come to the dump. Why did some bears not come to the dump, and which ones were they? We knew that the bears came ashore about the end of July or early August but that few were seen in Churchill before early to mid-October. Yet the dump was available through the whole period. Why did bears that knew about it not come in to feed as soon as they came ashore? In an ecological sense, was the dump an important supplemental food source to the bears? In other words, did the bears that fed at the dump receive benefits that could be measured in terms of increased survival or reproduction? What did the bears outside the Churchill area do when they were not feeding? How were the different age and sex classes of polar bears distributed during the period they were on land?

The reproductive ecology of the female polar bears offered some especially interesting questions. The Owl River denning area was probably one of the largest in the world. Did females come back to the same denning area or even to the same part of the denning area to have subsequent litters? What sort of variation might exist in productivity or survival of young between years?

Another fascinating question revolved around how often mother polar bears could have litters.

In most areas of the Arctic, female polar bears keep their cubs for two and a half years, so that they are able to mate and produce new litters at most once every three years. However, through the first ten years in the Churchill area, we had captured quite a few yearlings by themselves, something almost unknown in other areas of the Arctic. This seemed to suggest that at least some of the females were weaning their cubs at only one and a half years of age and were thus capable of breeding every two years. If this was true, it represented a major departure from the normal pattern and indicated a new dimension in the reproductive biology of polar bears. (The results of the work on reproductive ecology done in the Churchill area are presented in the chapter on reproduction.)

As if all these fascinating questions were not enough reason to study polar bears at Churchill, there were also nonbiological considerations: easier logistics and reduced cost. Instead of being spread over thousands of square miles of ice, far from camps or fuel supplies, the whole population was ashore for four months during the ice-free period and concentrated within about 250 kilometers (160 mi.) of Churchill. Through most of that period, there is no snow on the ground, making the white polar bears easy to sight. Long days and warm temperatures through the summer and fall make outside work efficient and comfortable (except when the bugs are rampant). Churchill itself is less costly to travel to than most locations in the Arctic. Equipment and specimens can be shipped by rail, which is far more economical than by air freight. We have always received a tremendous amount of support, assistance, and encouragement from the Wildlife Branch of the Manitoba Department of Renewable Resources, and that has helped immeasurably. Last but not least were the people of Churchill, who had a deep interest in the polar bears and helped us in many ways. The combina-

tion of these factors made Churchill one of the best places in the world to study polar bears then and today.

Besides my group's work on behavior and ecology, fascinating studies were conducted by other scientists on polar bear physiology and on deterrents for repelling problem bears. Results of these studies are covered in the chapters "What Makes a Polar Bear Tick?" and "Conflicts between Polar Bears and Humans."

## Cape Churchill

The cape is an almost flat esker extending north into a sandspit that reaches out into the shallow tidal flats of Hudson Bay, about 49 kilometers (30 mi.) east of Churchill. It is located at the point where the east-west orientation of the coast suddenly bends southward again to form the western coast of Hudson Bay. The lack of geographic relief can give the impression of bleakness or isolation on an overcast, windy day as the roaring waves of Hudson Bay pound on the beaches. The coastal plain in this area is treeless; the four-foot-high willows around some of the small lakes are the tallest plants. The remaining vegetation consists largely of sedges, grasses, and a few flowers and mosses. In fact, the place can seem decidedly unremarkable except for the fact that it is not at all uncommon to get up on an October or November morning and see thirty or more polar bears. At that point, one's perspective changes dramatically and Cape Churchill becomes one of the world's wildlife jewels.

There are always a few bears on the cape after breakup in the summer, but it is later in the season that things become more spectacular. All through the fall, and especially just before freeze-up, more and more bears move out to the coast from inland areas to join those that have spent the summer there. Many of them begin to meander up to the cape. Although some subadults and family groups go to the cape late in the season, most of the bears there are large adult males.

We knew that few of the big adult males of Cape Churchill visited town but we did not know what they did all day while waiting for the ice to form again. We needed to be able to stay out on the cape with the bears and watch them to find out. However, camping in a tent on the ground, literally in the midst of a herd of polar bears, was an unattractive option. We decided we needed a tower. There were a couple of wooden towers between Churchill and the cape that were used for observing rockets fired from the rocket range east of town. The military commander said we could have them and he would have the soldiers move them for us as an excercise during winter maneuvers. However, it never happened.

We began scrounging again. This time the Manitoba Wildlife Branch came through. Dick Robertson, the Regional Biologist from The Pas, had an obsolete fire lookout tower dismantled. The first 45 feet of steel were railed to Churchill and then flown out to the cape in a Twin Otter. With the help of a couple of professional steel-riggers from Manitoba Hydro, we had a magnificent tower in a few days. On top we assembled a ten-foot-square observation hut prefabricated by Dale Cross of the Manitoba Wildlife Branch at The Pas. Concrete anchors were set in the tundra and the whole unit was guyed down with heavy cables as a precaution against the strong winds the area is famous for.

Paul Latour and Dennis Andriashek began work in the tower in October of 1976. The first night a rip-roaring blizzard screamed out of the northwest and shook the tower back and forth until well into the next day. As the night wore on, the storm developed into what later came to be known as a "hummer." When the wind becomes strong enough, the cables that guy the tower down begin to vibrate like guitar strings and set up an impressive but ominous hum. The two men were faced with a classic dilemma. They could stay in the untested tower through the pitch-black night as it swayed back and forth. Or, they could risk climbing down the exposed ladder in the storm (no mean feat in itself) to seek refuge on the ground in the willows where they knew the polar bears would also be hiding from the wind. Finally, they decided to trust that the steel-riggers knew what they were doing. They were right; more than ten years later, the tower is still standing.

After the constant noise of the helicopter and all the people present around the Churchill area, studying bears in the peaceful atmosphere of the cape was always a treat. Initially, the most important thing to do was just watch. We counted the bears at regular intervals and noted areas they occupied. We monitored individual animals for long periods of time and kept a continuous record, known as an activity budget, of everything they did. Paul logged over 2,500 bear hours of observation for his M.Sc. thesis.

We also needed to be able to recognize individual bears through the season or between years. So, when they came to the bottom of the tower to investigate the camp, we would quietly lean out of the window and dart them. The initial reaction of most bears was one of confusion. They did not know where the noise came from or what it was. Some walked away a few hundred yards before falling over, but quite a few just lay down at the base of the tower. They were tagged and measured and large numbers were painted on their sides so they could be recognized at a distance. Capturing and marking the animals was particularly important for Paul because he wanted to study the relationships between individual animals. It was very important to know both how old and how heavy individual bears were.

And what did the polar bears of Cape Churchill do all day? To a large degree, not much of anything. Adult and subadult males were inactive 79 and 74 percent of the time respectively, while

POLAR BEARS

subadult females were inactive 56 percent of the time. All bears spent very little time feeding, only 1.25 percent of the total. These results were quite similar to what Brian Knudsen found several years earlier when he observed polar bears spending the summer on North Twin Island in James Bay. Along the coast, grass and kelp washed up on the beach are the main food items but it appears that little is actually eaten. So, with a good supply of fat on their bodies, the bears lie about, expending the energy of their fat deposits as slowly as possible, until they can return to the sea ice after freeze-up to hunt seals again.

Yet within this seeming inactivity, order is still discernible. As the density of animals increases through the fall, there is segregation between the adult and subadult males, with the larger older animals tending to prefer the extremities of the sandspit. The subadults maintain a greater distance from the adult males than the adults do between themselves, which probably indicates a certain amount of nervousness on the part of the subadults.

The most interesting aspect of the active behavior of these bears was the time they spent in social interactions with each other; 3 to 4 percent of the total time in adult and subadult males and about 8 percent in subadult females. The proportion of time spent in social interactions seems high for a large carnivore that we generally think of as being solitary and generally avoiding other bears except in the breeding season.

Interactions are mainly between bears of the same age and sex class, although some also initiate contact with bears of other classes. Not surprisingly, subadults do not initiate many interactions with adults. Curiously, even though the fall is six months away from the breeding season, both adult and subadult males approached subadult females more often than any other group. That may be why subadult females were more active than any other group. Besides moving on

their own accord, they were being forced to respond to the activities of other bears as well.

The most interesting social behavior exhibited by the bears in the fall is the ritualized play fighting of the adult males (pp. 153–55). This behavior has been seen by many people and is much loved by photographers because it gives them much more action to record than a resting bear does. During this play fighting, the males come into body contact as if in serious combat, but without injuring each other. These ritualized bouts are three to four minutes in duration. A pair of thousand-pound males fighting is an awesome sight. Each is capable of inflicting terrible wounds on the other, and yet they do not. There are several repeated behavioral patterns such as mutually rearing up on the hind legs with the forelimbs partially folded into the body, mouth open and head angled down, but making no contact with the partner. Sometimes this is followed by mutual pushing with the forepaws on the neck or chest of the opponent in an attempt to force him down. Facial contact is common; the bears may touch noses and then one may rub the area of the neck just behind or below the head. Facial contact may suddenly give way to a bear lunging toward its partner and pushing him on the neck or chest with the forepaws. Inhibited bites take place, in which one male may grip an opponent's neck or shoulder by the teeth without taking advantage of its purchase to cause serious damage. Much of the jousting seems oriented toward trying to put the opponent off balance. But even when this is done and a male has pinned his opponent, the bites and pushes are still inhibited so as not to cause serious wounds. Bouts usually end with the loser walking or running away. Interestingly, it was the bears with whom fights were started that most often terminated the encounters by fleeing. This suggests polar bears are good at assessing the size and strength of potential opponents. In this situation at least, they initiate encounters with bears of a similar or slightly smaller size. Sometimes, a third huge male will stand and watch the protracted combat of two of his compatriots, refraining, like a gentleman, from joining in. Once the scrap is over, he may then challenge the winner but he does not join in punishing the loser as dogs sometimes do.

We do not fully understand the significance of the male polar bears' ritualized fighting in the autumn. We suppose that it provides the bears with experience in fighting and an opportunity to improve their motor skills and coordination in circumstances where the risk of serious injury is minimized. This could have survival value by helping males to learn to fight more effectively when competing for mates or food out on the sea ice.

## The Bears at the Dump

Capturing bears out at the cape from 1976 to 1978, we learned that large males rarely ever visited Churchill, although subadult males did. More interesting, though, we found we were regularly recapturing males that had first been caught at Churchill as cubs with their mothers, or lone subadults. Some of these repeaters were caught as often as three or four years in succession and then not seen again. Suddenly, we found we were recapturing them along the coast. It seemed everyone's attention had been focused on controlling conflicts and no one had given any thought to why so many of the tagged male bears were not being recaptured, while many of the adult females were. Clearly, it was time to study the ecological importance of the dump.

Nick Lunn got the job of conducting a pilot study of the behavior of the bears at the dump, a project he did very well and developed into his M.Sc. Through the fall from 1981 to 1983, Nick drove out at dawn, parked his truck a few hundred meters from the dump, and recorded everything the bears did until dark. After a while, he became almost as much of a tourist attraction as the bears themselves. Visitors and tour buses would stop regularly to get the latest information on what bears were around, what they were doing, and anything else that might be interesting.

Besides observing the bears in the dump, we tried to do comparative studies of the behavior and activity budgets of bears of the same age and sex classes that did not visit the dump. This was complicated by the fact that most of the families were far inland where study was difficult. Some observation was possible along the open coastal areas.

Every bear that came into the dump was captured as soon as possible after it was first seen, tagged, and painted with a large number on the side so it could be identified from a distance. At first, we painted both sides but soon changed to numbering one side only. This left a clean side for tourists to photograph and a side for science. At one point, the crew of bears around the dump, each sporting a large black number, became known as "the hockey team." At the end of the season, we recaptured a number of them to see if they had put on weight while feeding at the dump.

Between 1966 and 1983, 207 individual bears were captured at the dump. Sixty-seven (32 percent) were recaptured there at least once after the year of their original capture. Several adult females returned repeatedly over a period of ten or more years. One female (Linda), first caught there as a yearling in 1967, kept coming back with successive litters of cubs until 1985 when she was nineteen (see the chapter on reproduction). Another female, first caught at the dump as a cub with her mother, was recaptured there with her own cubs several years later. Up to 1983, thirty-three individual adult females had brought fifty-seven litters (totaling 101 cubs) to the dump. After weaning, 21 percent of the cubs were known to have returned to the dump. These examples clearly demonstrate the importance of cubs learning from their mothers. However, males that returned to the dump to feed did so only until they were about four to six years old and were rarely seen in the area again. This stands in marked contrast to the situation with black and brown bears, for adult males that have learned to feed in dumps continue to do so indefinitely.

Another difference between the polar bears in the dump and their black and brown bear cousins was in dominance. At the Churchill dump, adult females with cubs, not males, were the dominant animals. The few adult males that entered the dump kept out of their way. Usually there was a dominance hierarchy between the family groups as well, with some families staying well out of the way of others. One year there was a remarkable deviation from this pattern of mutual avoidance and occasional aggressive interaction. A ten-year-old female with twin female cubs of the year and an eighteen-year-old female with twin yearlings fed, walked to and from the dump area, and rested together for over a month. At no time were these females aggressive toward each other, although they were to other family groups. The four cubs sometimes played together while one mother watched over them and the other fed at the dump. We have never seen anything like this before or since. I have often wondered if the younger female was the daughter of the older one, which might explain their unusually relaxed manner with each other. Unfortunately, they were both first caught as independent adults, so the possibility of their being related remains a mystery.

When the polar bears first arrived in the dump each autumn, there was no significant difference between their weights and those of bears of similar age and sex classes that did not go to the dump. However, there was a marked difference by the end of the fall. Bears outside the dump lost 0.3 to 0.4 percent of their body weight per day, whereas most of those that fed at the dump gained 0.1 to 0.6 percent of their initial body weight per day. For an adult female nursing her cubs of the year away from the dump area, this meant an average loss of 34 kilograms (75 lb.); a similar female at the dump might gain as much.

Even so, this nutritional benefit did not translate into improved reproduction or survival. There was no significant difference in the litter size of females that fed regularly at the dump when compared to those that did not. It was more difficult to evaluate the survival of the cubs of dump females. A comparison between the rate at which they were recaptured and the recapture rate of cubs from elsewhere in the study area showed no differences. However, there did appear to be some detrimental side effects from feeding at the dump. For example, one bear was found dead, apparently poisoned by eating part of a car battery. Subadult males that feed in the dump, and later wander about town, may have lost some of their natural fear of humans. A significantly higher proportion of subadult males with a record of being in the dump have been shot as problem bears than have those without such a record. Similarly, becoming used to humans around the dump may make polar bears less wary of Inuit settlements along the coast to the north of Churchill. Of tagged bears shot by Inuk hunters along the western coast of Hudson Bay, a higher proportion of subadult males with a record of having been at the dump are shot than of subadult males that have not been recorded there.

These results indicate that feeding at the dump does not confer detectable advantages upon the bears that feed there, so why do they do it? The bears all come ashore in mid-summer, but those that go to the dump usually do not do so until early to mid-October, even though they obviously know it is there. Probably less than 5 percent, and certainly less than 10 percent, of the population have been recorded there in any one year. This also tends to downplay its overall importance.

It seems fairly clear that the strategy the bears have evolved is to come ashore with as much stored fat as they need to survive until freeze-up. They are as inactive as possible so as to conserve energy. However, some bears may not have laid down quite enough fat to last for the whole period. They become nutritionally stressed as their time on land wears on. Two groups of bears are most likely to feel this strain: females with cubs of the year or yearlings because they have the additional energy drain of producing milk to nurse them, and subadults that have not yet learned to hunt effectively. It is probably no coincidence that these two groups frequent the dump most commonly. The reason they do not appear until later in the fall is probably because that is when their fat reserves are getting low. The reason subadult males stop coming to the dump after they are about four or five years old is probably that they have become proficient enough at hunting to be able to deposit enough fat to see them comfortably through their time on land. Thus, although they almost certainly still know the dump is there, they no longer visit it. In addition, it seems that mature polar bear males are more circumspect about humans than bears of other age and sex classes, and do not usually come around inhabited areas.

The relationship between the amount of fat deposited for survival through the ice-free period and the occurrence of bears around Churchill was dramatically demonstrated in 1983. In that year, presumably because of some change in the distribution or abundance of seals out on the bay, the whole population of polar bears came ashore about 15 percent lighter than usual. Bears began to feed in the dump in early September, a month earlier than usual. More bears were recorded at the dump than in several previous years, and there were more human-bear conflicts than in the previous decade. It seems likely that by monitoring the weights of the bears as they come ashore in the summer, we might be able to predict the approximate level of the human-bear interactions in the Churchill area that fall.

The final conclusion of the study on the ecological importance of the dump to the polar bears was that, although some of the bears will scavenge on garbage given the opportunity, this be-

havior confers no benefits and is not necessary for the survival of the individuals that do it. The results of this study helped set the stage for the next phase of the bear management program: trying to break the cycle of bears returning to the dump each year. From the tagging and recapture data, it was obvious the biggest offenders were a relatively small number of adult females that kept teaching their cubs to feed there. Some subadults also discovered the dump on their own and then continued to return in subsequent years. If the bears that came to the dump received no reward (food) each time they returned, could the cycle be broken?

In 1981, the Manitoba Department of Natural Resources built a "polar bear jail" capable of holding at least sixteen individual bears and four family groups. Now, when problem bears come into town, they are immediately captured and locked up. They are held there until freeze-up when they are released back onto the sea ice. If the jail fills up, and more bears keep coming into town, the inmates are flown north along the coast 80 kilometers (50 mi.) or so and released, in order to make room for the next visitors.

The research showed that the captive polar bears, like their compatriots on the coast, did not need to eat. Thus, they are not fed while in the holding facility. Otherwise, feeding might provide a reward that might attract them as much as the dump itself. Adult females with long histories of returning to the dump are being removed from the population and sent to zoos. This breaks the chain through which successive litters of cubs learn to frequent the dump.

Although this program is expensive, it has largely succeeded in preventing polar bears from walking through Churchill, where they might be killed as a threat to human life and property. One consequence is that the dump is now a pretty quiet place. Although it will be several years before the effectiveness of the experiment can be evaluated, it remains one of the more creative programs being undertaken in polar bear management.

# Conflicts between Polar Bears and Humans

The history of human expansion into the wilderness has been one of continuous competition with large carnivores for space and resources. In earlier times, over the entire Northern Hemisphere, human appropriation of habitat for agriculture and ranching, or just plain fear of predators, drove species like the wolf and the brown bear into remnant populations and local extinction throughout much of their original range.

Polar bears, too, come into conflict with man around the edges of the the polar seas and arctic archipelagoes. Besides scavenging at sites of human habitation, be it at Inuit hunting camps, weather stations, or settlements, polar bears are more willing than other bears to prey upon humans. For example, few black bears ever attack humans. Although more humans are attacked by brown bears, a high proportion are mauled but still left alive; not a pleasant prospect, but better than being killed. However, a polar bear attack on a human ends only when one of them is dead.

When a polar bear attacks a human, the victim is often unaware of the bear's presence until it appears at close range. The circumstances can vary greatly but the element of surprise is the bottom line, as a couple of examples will illustrate.

On 1 July 1961, Tony Overton and three colleagues from the Canadian Department of Energy, Mines, and Resources were doing some seismic studies on the sea ice near the southwest coast of Ellef Ringnes Island. Normally, all four men slept in a large tent but one night, since one of the fellows had spilled liquid on the floor, the other three decided to sleep outside on the sea ice since the weather was mild. They had not seen any signs of bears, or even of seals, during the time they had been camped there. Even so, Tony put his loaded rifle on the ice by his right hand before crawling into his sleeping bag and going to sleep. To this day he doesn't know why, but for some reason he suddenly woke up at about four o'clock in the morning and saw a polar bear approaching about 65 meters (200 ft.) away. He reached for his rifle but before he could lift it, the bear charged. It seized him by the arm as he was raising it over his head in a last-minute act of self-defense. The bear dragged him out of his sleeping bag and about 15 meters (50 ft.) across the ice. Tony desperately shouted at his slumbering colleagues three times before finally managing to wake the late Bill Tyrlik. Bill quickly assessed the situation and swept up the rifle to shoot. By that time, the bear had dropped Tony and was charging him instead. Bill did not even have time to aim before the bear reached him, so he shot from the hip. He dropped it, with one lucky shot between the eyes, on the end of his sleeping bag. (One of Tony's other companions commented some time later that he had heard all three of his yells but thought he was just having a nightmare!) It seems likely the bear thought they looked like seals on the ice from a distance, but when it seized Tony it knew it had something different. Maybe it was the fact he held his arm up in the bear's face, or possibly his unfamiliar scent. As I have told Tony since, he should be glad the bear didn't seem to think he was a seal because the first thing they do after catching one is chew on its head and neck to make sure it is dead.

On another occasion, Tom Smith and the inimitable Inuk hunter Jimmy Memorana had just set up their tent on the sea ice one cold evening in early April. They were unpacking inside and Tom was kneeling over his gear doing something when Jimmy matter-of-factly said, "Don't move." Jimmy does not fool around, so Tom froze. In a single motion, Jimmy swung his rifle over Tom's back and pulled the trigger, killing a polar bear with a single shot just as it was starting to come through the back wall of the tent.

In two other instances, a large polar bear stalked a person to within a few feet in some pressure ridges. In both cases, when it was within a few feet, it let out a slight puff of air just before attacking. One was the famous arctic explorer Vilhjalmur Stefansson, and the other was Jimmy. Both had been around polar bears enough that their reflexes took over and they were able to save their lives by shooting the bear. Had the bear not given a warning, neither man would have lived to tell of the experience.

Having watched so many polar bears hunting, I am intrigued most by the momentary pause to blow out some air. I doubt polar bears do that when they hunt seals and I think it suggests they know they are hunting something different, and possibly a little dangerous. It is interesting to think that even a polar bear gets a little nervous at times. (And so he might be when hunting someone like Jimmy!) Even so, the point of these stories is that one always has to be alert in polar bear country.

## Identification of a Problem

For a number of years, wildlife agencies have generally tried to discourage the killing of so-called problem bears of all species unless it was absolutely necessary. Even so, few people actually worried much about those that did get shot. With polar bears, it was often difficult to obtain accurate figures on how many bears were being shot in "defense of life and property" (such cases are called DLPs). An unknown number of them were killed when they came into settlements and were simply included in the quota.

Ray Schweinsburg, formerly the polar bear biologist for the government of the Northwest Territories, worried about it, and rightly so. He frequently pointed out that the annual quotas for Inuit hunters were already set near the limit of the estimated sustainable yield for each subpopulation. This meant that an unregulated kill of "problem bears" was additive and might make some populations vulnerable to being overharvested. A few numbers make his point. In 1977, the recorded kill of problem polar bears in the Northwest Territories was ten. Five years later, it was over forty. The number of additional problem bear kills in the rest of Canada would likely have been another twenty or so per year but there was no way of knowing the total. Similarly, we knew that some problem bears were being killed in Alaska, but because the Marine Mam-

The famous female, Linda, rests near Churchill with two cubs.

Twenty-six-year-old Linda as a resident of the Rio Grande Zoo in Albuquerque, New Mexico

Collecting blood samples for physiological studies of health and determining pregnancy

Measuring the girth so the weight of the bear can be determined

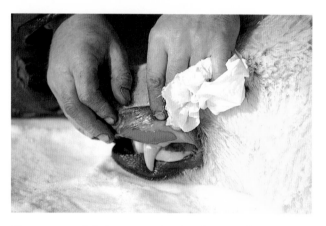

Green tattoo ink is put on the inside upper lip.

It is stamped with an individual number with tattoo pliers to give the bear a permanent identity.

This tattooed number is several years old.

Checking the teeth of an adult male for wear. (Photograph © Ian Stirling.)

Andy Derocher passes cubs from their maternity den to Steve Miller for measurements and tagging.

Ian Stirling weighs a drugged cub after putting tags in its ears. The cub recovered a few minutes later and lay down beside its mother.

Problem bears are sometimes immobilized and airlifted several kilometers away from town where they are released unharmed.

A superfat pregnant female weighing almost 500 kilograms (1,100 lb.) has a radio glued to the top of her head so scientists can find her den in the winter. The radio will fall off in the following spring when she goes through the annual shedding of her hair. (Photograph © Ian Stirling.)

An ancient adult male reduced to skin and bones by old age and starvation

Some males are still rolling in fat after fasting for three months.

Even flames do not deter polar bears from feeding at the Churchill dump.

Churchill, east of the airport

An example of habitat modification. The cracks and open water that form near the *Molikpaq* attract seals. Polar bears may remain at the site hunting for days or even weeks.

Viewing polar bears from the Tundra Buggy

Polar bears investigate the Tundra Buggy.

A large bear bites a window 3 meters (10 ft.) above the ground. Even when one is in a vehicle, a bear can be dangerous.

mal Protection Act placed restrictions on the ability of government agencies to monitor the harvest taken by native people, it was difficult until recently to keep tabs on this aspect.

Nor was the problem just a North American one. At the 1981 meeting of the IUCN Polar Bear Specialists Group in Oslo, Norway, a very interesting perspective came out. The Soviet Union had given the polar bear protected status in 1955 and Norway did the same thing at Svalbard in 1973. At the time, the populations were probably depleted in both areas and, as a result of being hunted, the bears likely also had a healthy respect for people. Initially, few bears were seen, but then things changed. Because of the lack of hunting, the polar bear populations probably grew larger and their fear of man grew smaller. After only five or six years, the weather stations and settlements in Svalbard began reporting increasing numbers of polar bears in the winter. Soon, half a dozen bears were being shot each year. At the 1981 meeting in Oslo, Savva Uspenski showed us some incredible photographs of polar bears wandering about in small villages on the northern Siberian coast. In one, a lady was standing on her doorstep calmly feeding a large male. Uspenski said the bears were becoming quite brave, which raised concern among the Soviet authorities. Well they should be if his photos were any indication of what was going on.

These parallel experiences with polar bears in Svalbard, the Soviet Union, and to some degree, Churchill also serve to underscore an especially important point when it comes to the conservation of large carnivores. Just stopping all killing of these animals is not a guaranteed means to save the species, as preservationists would like to think. Polar bears, like other large carnivores, are dangerous animals. Although most individuals are remarkably good-natured and will avoid inhabited areas given half a chance, they do sometimes kill people and cause serious damage to property. Consequently, those responsible for

their management must have a variety of options open to them. Those options must include the authority to remove individuals when the occasion demands it.

## What Bears Cause Problems?

Before you can solve a problem you have to define it. Which polar bears cause problems and why do they do so? At Churchill, Nick Lunn showed that most of the bears that came to the dump were subadult males or females with cubs. In general, both groups had a tendency to be thinner and hungrier. In years when the bears came ashore lighter than usual, more bears showed up at the dump. The females with cubs tended to remain there. However, subadult males tended to wander more, and several of those that wandered into town were shot.

From 1973 to 1983, twenty-eight bears were reported killed as DLPs in the eastern Beaufort Sea. This is a minimum number because, for at least the first half of that period, problem kills in the villages and hunting camps were poorly documented and specimens were not always obtained. However, of the sixteen bears whose ages were determined, twelve were five years of age or less and two were cubs of the year accompanying their mothers. The average age of the twelve subadults was 2.25 years. This pattern in the eastern Beaufort Sea, of the greatest number of problem bears being subadults, is similar to what has been found in other areas. The bears that attacked Tony Overton and tried to enter Tom Smith's tent were also subadults. These are the bears that have the most difficult time hunting. They have just been turned out on their own by their mothers and they have not yet become proficient hunters. When they do kill a seal, they also have a higher risk of having it taken away from them by larger, more dominant bears. Scavenging is an important part of their strategy for survival. Consequently, subadults are more likely to be in poor physical condition during periods when food is less abundant. There may be little option for a small number of them but to try scavenging on garbage, and occasionally even preying on humans, in order to survive.

This was brought home rather forcibly in the eastern Beaufort Sea in the winter of 1974–75. The size of the seal populations had been greatly reduced in the previous summer. It seemed logical that if there were fewer seals, subadults might have an even more difficult time and, consequently, be in poorer condition. We predicted an increase in defense kills. Sure enough, there were seven problem bears killed that winter, compared to one or two in normal winters. One of those bears killed a human at an oil company camp in January 1975. The seal populations remained low again the following summer and another seven problem bears were shot in the following winter, although no one was killed. After that, the seal populations recovered and the number of problem bears dropped back to one or two a year.

At Churchill, we also noticed that in some years, all the bears weighed less than usual when they came ashore to fast through the open-water period in the late summer and fall. In those years, more females with cubs came into the dump to feed late in the fall when their stored fat reserves ran low. In fact, one of the problem bears killed in the Beaufort Sea was an adult female accompanied by cubs of the year. In the High Arctic, adult females with cubs twice featured in problem bear incidents during the winter; both were very thin. In one case, the bear dragged an oil worker away into the dark polar night. Another man had to follow the blood trail in a front-end loader before he could drive the marauding bear away. Twice in recent years, thin adult females with cubs were shot while trying to break into occupied houses at railway maintenance camps east of the Churchill denning area.

With polar bears, the common thread in problem incidents is that hungry, sometimes starving, animals have nothing to lose by trying something different. After all, investigating new possibilities is the hallmark of success for carnivores.

While we know that nutritionally stressed subadults, and females with cubs, enter human habitations, we do not know what age and sex classes of polar bears are most likely to threaten humans out on the ice in prime polar bear habitat. Would other age and sex classes become involved? For example, few adult male polar bears become problems at dumps, which is just the opposite of the case with black or grizzly bears. But, would they be more aggressive toward humans out on the sea ice?

My impression is that adult males have learned to be wary of people and are less likely to be a threat. At Cape Churchill, when we hung out a smelly bait specifically to attract bears so we could tag them, we consistently noticed it was the largest and fattest males that were the most cautious about approaching. Similarly, Savva Uspenski told me that along the Siberian coast, the large males were the most wary of all the age and sex classes, and usually only came into the villages in the middle of the night. Over the years, our consistent impression has been that the bears in the best condition overall were the adult males. Since they experience the smallest amount of nutritional stress, they probably investigate human habitations least often.

## The Human Factor

An important aspect to remember when considering human-bear conflicts is that some of them are not the fault of the bear. Feeding polar bears and otherwise acclimating them to close human presence sets the scene for problems. Bears are individuals and they learn quickly. They are adept at spotting opportunities for a meal while not giving any indication of it. Two incidents illustrate this.

One autumn, I allowed a photographer and his friend to stay in my research tower at Cape Churchill for a couple of weeks when we were not occupying it. There are some cross struts about 4 meters (12 ft.) above the ground. They put some boards across these so they could stand lower down to get better pictures. Off to one side, under the tower itself, there were also a couple of fuel drums. A large bear was walking about below, apparently not paying attention to anything in particular. Then, like greased lightning, it leaped onto the drums and used them to spring up at the man. He got a claw into the lower pantleg and tore it open before falling to the ground, leaving the careless person a little wiser. The bear could not have jumped the full 4 meters, but he figured out that he could reach him if he got to the top of the drums first.

In another incident out at Cape Churchill, an experienced polar bear photographer got a little careless. He was taking pictures from the Tundra Buggy, an all-terrain vehicle rather like a high-rise bus mounted on huge tires. A large, very thin bear had been hanging around the Tundra Buggy for a few days and had disappeared from view beneath it. A rare ivory gull landed nearby, causing great excitement among the photographers. The man hung his arm out the window to support his camera as he photographed the bird. Seeing his opportunity, the bear quickly emerged from under the vehicle, stood up on his hind legs, and seized the man's arm in his jaws. The tour guide pounded on the bear's head hard enough to make him let go but not before the photographer's arm was badly ripped open. A dramatic rescue operation across the tundra to the hospital in Churchill saved his arm and his life.

In 1983, the polar bears along the coast south of Churchill came ashore much lighter than usual for some unknown reason. This meant they had less stored fat to live on until they could re-turn to the ice. Consequently, bears began to get thin by late summer and decided to check the Churchill dump and town site. That fall, the bear patrol recorded one of the highest numbers of problem bear calls ever.

Everyone knew there were lots of bears around. Yet, a local man went to scavenge in the remains of a burned-down hotel late one night after leaving the bar. He stuffed his pockets full of meat and other food from the abandoned freezer and was apparently walking off in the dark when a bear attacked him from behind. The man was killed and the bear dragged his corpse down the main street, where the bear was finally shot. The townspeople were outraged and the press had a field day with headlines about a northern town under siege by polar bears. As unfortunate as both these incidents were, the human in each case was more at fault than the bear. The bears were simply behaving naturally. Someone who places himself in a polar bear's mouth, so to speak, has only himself to blame if he is bitten. In polar bear country, it pays to observe the Manitoba motto: "A Safe Bear Is a Distant Bear."

## Detection and Deterrents

In 1981, the Government of the Northwest Territories hired Gordon Stenhouse as its first bear deterrent biologist. His three-part assignment was impressive: to develop methods of detecting bears (not just polar bears but grizzly and black bears as well) before they got into camps or other inhabited areas; to develop methods of chasing the bears away without hurting them; and to develop an educational program on bear safety for people working in bear country.

The first two goals are straightforward enough. One of the main reasons a polar bear in camp is so dangerous is that people are not likely to be prepared when they encounter it. If you have a warning system, then people can be alerted and take steps to try to scare it off. The real problem is how to scare it off once you know it is there, something that is often more easily said than done.

To test detection and deterrent systems, you have to go where there are lots of bears. It would be hard to find a better place than Cape Churchill in the fall. So Gordon went out to our observation tower there and turned it into something resembling a combat zone. There were speakers to broadcast loud noises, including the sounds of barking dogs, and a siren. Microwave detection units and a trip wire system were set up to detect bears, and an electrified barbed wire fence surrounded the compound to prevent the bears from entering. There was an impressive assemblage of noise-making devices, such as cracker shells that were originally developed for frightening waterfowl from crops. He had a 38-millimeter riot control gun that fired rubber bullets. Finally, on the roof of the hut at the top of the tower, there was a set of floodlights so he could observe bears at night.

Some bears approached of their own accord while others were specifically attracted with pieces of seal or whale blubber. The reason for providing fat was to give the bear the maximum incentive to remain. This would test the effectiveness of the various deterrents more rigorously.

Some of the results were as impressive as the appearance of the camp. In 1982, both the microwave and trip wire systems were 100 percent effective in detecting the approach of 187 bears. However, the automatically activated sirens did not deter the bears from continuing their approach. Eighty-seven percent of the approaching bears were quite unimpressed by the recordings of barking dogs. Ninety-three percent of the 42 animals tested went right through the electrified fence unaffected. Apparently it did not ground properly on the snow and frozen ground. However, another electrified fence used nearby on wet

ground during the summer grounded well and was quite effective.

Personally, I was especially interested in how the deterrents worked because that approach was obviously the key to saving the lives of many polar bears. Earlier studies on a variety of species showed that loud noises used alone were not very effective at frightening animals away because the animals quickly got used to the sounds and ignored them. That turned out to be the case with polar bears as well. Some of the bears were frightened by cracker shells or gunshots, but most quickly learned not to be concerned. As children are wont to say, "Sticks and stones will break my bones, but names will never hurt me."

What the polar bears needed was a loud noise coupled with what science calls a "negative stimulus." That means the bear had to experience something unpleasant, maybe temporarily painful, but not harmful. The sudden impact of a rubber bullet half the size of a pound of butter, fired from the 38-millimeter riot control gun at a distance of 40–60 meters, served that purpose admirably. There was no ignoring the smack of a fast-moving rubber baton on the backside and the results were unequivocal. All of the 404 bears hit with rubber bullets from 1981 through 1983 fled immediately, except for one. He was a large but very thin adult male who still would not move after he had been hit five times. This degree of determination illustrates why a starving bear is so dangerous. His principal concern is to find something to eat. He will risk death to get it.

Some of the bears came back a second or even a third time, but the "negative stimulus" of the riot gun sent them scurrying once again. The indication was that they had learned quickly and were not about to stay for more.

Although testing continued at night with the aid of the spotlights, a few bears learned to wait until the biologists went to bed. Then they investigated the area under the cover of darkness. The

POLAR BEARS

speed with which polar bears can assess and adjust to new situations should never be underestimated.

A disadvantage of the riot control gun is that, in Canada at least, it is classified as a restricted weapon. Only a peace officer is legally allowed to carry and use one. Consequently, one cannot be kept in every camp where bears of any species might be encountered. So, although the principle was excellent, the application needed to be modified. Peter Clarkson then brought his enthusiasm to the project. He began developing a plastic slug that could be fired from the universally available 12-gauge shotgun. The project is still developing but the initial results look quite promising.

Clarkson also developed a first-rate educational course with an accompanying manual called "Safety in Bear Country." He ran a series of workshops to teach people about bears, how to detect them, and how to deter them without killing them. Copies of the manual have now gone to all the circumpolar countries. Although there is still much to be learned, this project on detection and deterrents has been one of the most important initiatives taken on the conservation of polar bears to date.

# Conservation and Environmental Concerns

Polar bears were hunted by aboriginal peoples throughout the polar basin for hundreds, if not thousands, of years before the advent of modern civilization and sophisticated weapons. Since the indigenous people possessed only primitive weapons, the tactical advantage lay with the polar bear. Most of the bears killed were probably swimming in the water or resting in maternity dens. Even so, there are still many vivid stories in Inuit oral history of bears being killed on the open ice by incredibly brave men armed only with spears and aided by their dogs. However, unlimited hunting with stone age technology did not threaten the survival of the polar bear.

As firearms became more advanced and widely distributed, greater numbers of polar bears were shot. For example, the journals of the polar explorers of the nineteenth century report surprising numbers of bears being killed by most expeditions.

Commercial harvesting of polar bears was not widespread in most of the North American Arctic in the eighteenth or nineteenth centuries. Some trading took place in Alaska, and at Hudson Bay trading posts in the Canadian Arctic, but the harvest was not particularly profitable. The hides were heavy to transport and not worth very much compared to other furs. The majority of the polar bears were probably killed for clothing, and food for people and dogs.

In the European Arctic, however, fairly large numbers of polar bears were killed during the same period. Russian hunters and trappers were active in Svalbard by the seventeenth century and several trading ships sailed there each year. In the winter of 1784–85, a single Russian crew at Magdalena Bay killed 150 bears, and similar less well documented harvests were taken in subsequent winters. After reviewing all the available documentation, Uspenski suggested that the annual harvest could not have been less than 150 to 200 throughout the entire eighteenth century. In the nineteenth century, the volume of the Russian fur trade decreased, but Norwegian activity increased at the same time. Estimates based on admittedly incomplete information indicate that between 1875 and 1892, the annual Norwegian harvest at Svalbard was 144 polar bears, increasing to 415 between 1893 and 1908. The kill averaged 355 per year from 1924 through 1939 and dropped to 324 per year after World War II, from 1945 to 1970. Altogether, over 22,000 polar bears were killed during this period.

Uspenski summarized the annual harvests in the archipelago of Franz Josef Land as follows: in the 1800s, not fewer than 100; from 1890 to 1909, 100–150; from 1910 to the end of the 1930s, 150–200; and in the 1940s and early 1950s, not fewer than 50.

There is a long history of marine mammal hunting in the area of Novaya Zemlya. There are records of polar bear hides being exported as early as 1556. In the winter of 1835, there were eighty vessels with over 1,000 hunters on board. They were mainly hunting walruses, but most boats took polar bears whenever they got the opportunity. In most other areas, polar bears were taken regularly but in lower numbers, although the accuracy of the records is not good.

Interest in harvesting polar bears increased again in the first half of the twentieth century. In the Soviet Union, bears were taken all across the north at weather stations and hunting camps. Overall, Uspenski estimates that more than 150,000 polar bears have been killed or captured in Eurasia since the beginning of the eighteenth century—quite an impressive record. Most of these (60 to 65 percent) were taken in the western parts of the Barents Sea including the islands of Svalbard. Twenty to 25 percent were killed in the Chukchi Sea, and only 10 to 20 percent in the Kara, Laptev, East Siberian, and Bering seas.

## International Concern about Polar Bears

Through the 1950s and particularly during the 1960s, the rapidly increasing value of polar bear hides in North America and Europe, coupled with the increasing use of oversnow machines, stimulated unprecedented increases in the numbers of polar bears reported killed. For example, in Alaska, the trophy kill alone increased from 139 in 1961 to 399 in 1966. In Canada, between 1953 and 1964, the recorded harvest fluctuated between 350 and 550, while in 1967 it suddenly jumped to 726. The records are incomplete in most countries, so we will never know the actual numbers of bears killed.

Unfortunately, despite the large numbers of animals being killed, the countries in which the harvesting was taking place had little accurate knowledge of the size and movements of their polar bear populations. For example, one fairly widely held view maintained that all polar bears were part of a single circumpolar population, ranging at will from country to country throughout the Arctic. If this theory was true, the harvesting of bears by the hunters of one country could have had a significant affect on bear populations in other countries.

The first preliminary estimates of the total world population by both Canadian and Soviet scientists, albeit based on weak information, were as low as 5,000 to 10,000. Subsequently, Uspenski and Shilnikov estimated a world population of 10,000 animals based on a sample of only 58 bears seen on aerial survey flights made over the sea ice north of the Soviet Union in 1962, 1967, and 1968. Even though the scientific basis of these estimates was quite weak, there was no other information available with which to confirm or reject them. But if they were even partly true, the future of the polar bear could have been in jeopardy.

Worldwide attention to the plight of the polar bear was intensified by the public outrage over the way in which some polar bears were being legally hunted. The largest outcry was over the use of aircraft for hunting polar bears in Alaska. It was condemned for being extremely unsporting. Similar in its lack of sportsmanship, but less frequent in occurrence and considerably less publicized, was the shooting of polar bears from ships in the pack ice by tourists on board Norwegian tour ships near Svalbard. Some of these bears were helplessly swimming in the water when shot. Also on Svalbard, many of the polar bears taken by Norwegian trappers were killed at unmanned trap sites. A bait was attached to the trigger of a high-powered firearm aimed just above the lure. When a polar bear came along and pulled the bait, it shot itself. These "set-guns" as they were called, indiscriminately killed or wounded any bear that set them off, regardless of age or sex class.

## Events Leading to the Coordination of International Concern

In response to growing concern about the number of polar bears being killed each year, the first international meeting to discuss conservation of polar bears was held in Fairbanks, Alaska, in 1965. Representatives of all five "polar bear countries" (Canada, Denmark, Norway, the United States of America, and the Union of Soviet Socialist Republics) attended.

After much discussion, agreement was reached on the following points.

1. The polar bear is an international circumpolar resource.

2. Each country should take whatever steps are necessary to conserve the polar bear until the results of more precise research findings can be applied.

3. Cubs, and females accompanied by cubs, should be protected throughout the year.

4. Each nation should, to the best of its ability, conduct a research program on polar bears within its territory.

5. Each nation should exchange information freely and the International Union for the Conservation of Nature (IUCN) should function to facilitate such exchange.

6. Further international meetings should be called when urgent problems or new scientific information warrants international consideration.

7. The results of the First International Scientific Meeting on the Polar Bear should be published.

The IUCN was asked by the five nations to act as an information center and to coordinate the exchange of research results on polar bears. It was the Conservation Foundation in the United States that became the catalyst for the next stages of development. In 1967, the foundation funded Dr. Richard Cooley of the University of Washington to organize another meeting of polar bear scientists. He met with key persons in each of the five polar bear nations and at the IUCN, which subsequently hosted the first meeting of polar bear specialists at its headquarters in Switzerland.

Travel funds for an invited group of scientists were provided by the Conservation Foundation. Besides discussing a number of scientific questions and management needs, the scientists also organized themselves into what is now the IUCN Polar Bear Specialists Group of the Survival Services Commission. The group met every two years, from 1968 to 1972, to discuss the coordination of research and managment of polar bears. Most important, they negotiated the objectives and text of the International Agreement on the Conservation of Polar Bears and Their Habitat.

## The International Agreement on Polar Bear Conservation

On 15 November 1973, the International Agreement on the Conservation of Polar Bears and Their Habitat was signed in Oslo, Norway (see the Appendix for the complete text). It was ratified by three countries (the minimum number required) on 26 February 1976 and came into effect three months later. The remaining two countries ratified shortly thereafter, making support for the Agreement unanimous. After completion of the trial period of five years, the Agreement was unanimously reaffirmed for an indefinite period in 1981. (The proceedings of all the meetings of the Polar Bear Specialists Group are available from IUCN, Gland, Switzerland.)

From a political point of view, the most significant aspect of this remarkable Agreement is that it represents the first time the five arctic nations cooperated together to negotiate a framework for resolving a unique circumpolar concern. There is still no other polar subject upon which the circumpolar nations have come to mutual agreement.

Equally significant, from a biological perspective, is that the Agreement is sound scientifically. It is not simply a protectionist document, which would have contributed little of substance. Although the Agreement allows the taking of polar bears (which includes hunting and capturing), it sets out specific conditions under which that may take place. These are: for bona fide scientific purposes, to prevent serious disturbance of the management of other resources, by local people exercising traditional rights, and for protection of life and property. All the countries agreed to conduct national research programs on polar bears, with particular emphasis on the conservation and management of the species and the exchange of data from those studies.

From an ecological point of view, Article II may be the most profound part of the Agreement. It states "Each Contracting Party shall take appropriate action *to protect the ecosystems of which polar bears are a part* [my emphasis], with special attention to habitat components such as denning and feeding sites and migration patterns, and shall manage polar bear populations in accordance with sound conservation practices based on the best scientific data." Clearly the intent is to protect polar bear populations from becoming endangered from overhunting or the detrimental effects of humans on the environment. Equally clear is the recognition that the conservation of a large and potentially dangerous carnivore requires a flexible management plan.

## Responses to International Concern

In the years following the first international meeting in 1965, the size of the polar bear kill continued to increase. The countries involved could not afford to wait until the results of long-term research studies were completed, or an international agreement was negotiated, before taking action. Recognition of the extent of international concern, and the obvious lack of information about the bears, provided sufficient impetus for action.

In 1968, the Northwest Territories of Canada imposed quotas on all the villages for the first time. In the absence of scientific information on the polar bear populations, the average of the previous three years' harvest was calculated separately for each settlement and a slightly lower value was set as the quota. It was explained to the Inuit hunters that this was an interim measure. Later, quotas would be revised, up or down, in response to population studies when they were eventually completed.

Soon afterward, two important committees were formed in Canada that meet annually and have made polar bears one of the best researched and managed species in the Canadian Arctic. The Polar Bear Technical Committee is made up of the research biologists from each jurisdiction that has responsibility for managing polar bears (four Provinces, two Territories, and the Federal Government) and invited experts in areas such as population modeling, physiology, and so on. They meet once a year to discuss all the research that has been done in the past year, plan new research, and consider the application of new research results to management and conservation goals. The Polar Bear Administrative Committee then meets to discuss the recommendations of the Technical Committee and to make decisions on the management of polar bears on a nation-wide basis. In this way, there is little time lost before new research results can be directly applied.

In 1971, Alaska ceased the unlimited bag once allowed residents for their own use and established an annual limit of three bears per person. The number of sport-hunting permits was limited in 1971 to 210 for the western area and 90 for the northern area. In 1972, the United States passed the Marine Mammal Protection Act and all hunting of polar bears ceased, except by native people for subsistence purposes. Ironically, because the Act provided no restrictions on subsistence hunters, there was no longer a closed season or protection for bears in dens or females with cubs. Although the total kill was reduced in Alaska, in terms of legal protection for polar bears, the United States went backward. Since then, a significant proportion of the kill has been concentrated on the most valuable portion of the population, the reproductive females. This is still an important problem that has not been completely resolved.

Meanwhile, similar limits were established in the European Arctic. In Svalbard, the shooting of polar bear cubs and females was prohibited in 1965. In 1967, it became illegal to use oversnow machines, boats, or aircraft to pursue or kill bears. In 1970, the number of permits issued for killing polar bears on Svalbard was limited to 300. These were divided between residents, trappers, weather crews, sealers, and tourist hunters. Finally, in 1973, the Norwegian government established a five-year moratorium on the hunting of polar bears in Svalbard. To date, this moratorium has not been lifted.

Following the 1965 meeting in Alaska, there were no changes in polar bear harvest practices in Greenland. The Inuit hunters there had been harvesting 100 to 150 polar bears annually for many years, with no apparent signs of overharvest, so

they continued to do so. The USSR had already declared complete protection of polar bears in 1955.

## Legal Protection of Polar Bear Habitat

Largely stimulated by the negotiation of the Polar Bear Agreement, there was a great burst of new studies in the late 1960s and early 1970s. As data accumulated, several areas of critical habitat for polar bears were given some degree of legal protection. Approximately 40 percent of the land area on Svalbard was protected by Royal Decree in June, 1973. This included three National Parks, two Nature Reserves, and fifteen bird sanctuaries. In 1976, the Northeast Svalbard Nature Reserve was made into a biosphere reserve under UNESCO's Man and the Biosphere program. Because of this, most of the denning areas and important summer sanctuaries in that region are now completely protected. Even entry by scientists to do research on polar bears is allowed by permit only.

The Northeast Greenland National Park, which includes a lot of good polar bear habitat, was established in 1973. People from neighboring settlements are allowed to hunt there but only as far as a sledge can travel into the park and return in one day. This gives the Inuit access to a traditional area but reduces the hunting pressure on the polar bears. Part of Melville Bay in Northwest Greenland is proposed as a reserve area where hunting and all other travel would be totally prohibited. All identified denning areas in Greenland now receive protection up to 12 miles out to sea.

In 1976, Wrangel and Herald islands, which have some of the most important polar bear denning areas in the USSR, were designated as State Reserves. Managers of reserves can stop or restrict all human activity within their boundaries, including research, and visitors are not allowed.

In Canada, Polar Bear Provincial Park was established along the Hudson Bay coast of northeastern Ontario. It is a wilderness area that contains important maternity denning habitat in winter and summer sanctuaries for bears of all age and sex classes during the open-water period in summer. No motorized transport is permitted there. In Manitoba, the Cape Tatnum and Cape Churchill Wildlife Managment Areas were established in 1968 to give managers the ability to regulate activities, including research, along the coast. The areas are fairly large and include most of the important denning areas and summer sanctuaries along the western coast of Hudson Bay.

In southeastern Baffin Island, Auyuittuq National Park, primarily established for its spectacular scenery, contains a small amount of polar bear denning and summering habitat in the fiords in the northern areas. New National Parks are also being considered that would include some polar bear habitat in the Canadian Arctic. On Bathurst Island in the Canadian High Arctic, Polar Bear Pass National Wildlife Area was established in 1986 to protect wildlife and habitat. It is not a high-density polar bear area but some animals migrate through the pass seasonally and a few females den there.

## Comment on the International Polar Bear Agreement

In the twenty-plus years that have passed since the first international meeting on the conservation of polar bears was held in Alaska in 1965, there has been impressive progress. Even though there are large gaps in our knowledge of polar bears, an enormous amount of research has been completed. Much of this information has already been translated into changes in regulations, and the protection of critical areas of polar bear habitat, throughout the Arctic. Because of the unique degree of cooperation between the circumpolar

nations, the International Agreement on the Conservation of Polar Bears and Their Habitat and the IUCN Polar Bear Specialists Group exist to promote research and coordinate their management. In an age when new species are becoming endangered at regular intervals, and environmental degradation continues on a global scale, the history of research and management of polar bears represents an international success story in conservation.

In his recent book, International Wildlife Law, Simon Lyster noted, "The Agreement has proved very successful as a legal conservation instrument . . . and . . . has undoubtedly contributed to the establishment of protected areas for bears, to restrictions on hunting, and to the substantial amount of scientific research that has been carried out in recent years." He also points out the weaknesses of the Agreement: the terms are not enforceable in any country and there is no infrastructure to oversee compliance. So far, just the existence of the Agreement has been enough to encourage countries to continue to honor its terms. However, as Lyster goes on to note, "The fact that Parties are not required to hold regular meetings to recommend ways of making the Agreement more effective has not yet been a serious hindrance, but it may make it easier for Parties to ignore the provisions of the Agreement if they prove to be a serious stumbling block to future industrial development in the Arctic."

## Environmental Threats to Polar Bears

By the 1970s, interest in offshore exploration for oil began to increase in several areas of the Arctic, including the Beaufort and Bering seas, the western coast of Greenland, Svalbard, the Canadian High Arctic Islands, and Hudson Bay to name just a few. Elaborate plans were and are being developed, for year-round drilling in ice covered waters, large-scale production fields, year-round shipping in special ice-reinforced tankers, interisland and subsea pipelines to take oil from the Arctic to southern markets. In addition, a whole network of new support facilities, aircraft, and settlements will be needed to support these acitivities. Offshore exploration and production activities have given rise to a whole new range of concerns for the welfare of polar bears and other arctic marine mammals. These include possible detrimental effects on polar bear habitat, damage to key parts of the food chain (seals in particular), and direct injury or death to the animals themselves.

In general, polar bears prefer to stay on the sea ice so that they can continue to hunt. Thus, exploration activities that take place during the open-water period are of little consequence unless some problem, such as a subsea blowout of oil, continued into the winter. If this happened, polar bears could come into contact with the oil when the area became covered with sea ice and they returned to hunt seals there. In most arctic areas, the greatest potential problems are likely to occur between freeze-up in the fall and break-up in the spring, in areas where prime polar bear habitat overlaps with proposed or extant offshore drilling and oil-production sites.

An example is the eastern Beaufort Sea, where the preferred seal hunting habitat for polar bears is characterized by the system of shore leads that run parallel to the coastline over water depths of about 20 to 50 meters (60 to 150 ft.). Besides being the favorite hunting habitat, these lead systems are also the main seasonal migration route for polar bears moving back and forth between their summering areas and winter hunting habitat. Individual bears may travel up to several hundred kilometers back and forth along the lead system within or between years. Consequently, a significant portion of the population is likely to pass near any particular point during the course of a winter. Unfortunately, this area of prime polar bear habitat overlaps significantly the area

that apparently has the greatest potential for oil production (see Fig. 8).

Although the chances of the "worst-case scenario" occurring are not great, an uncontrolled oil blowout under the ice during the winter could pollute a prime feeding habitat for the polar bears. The currents would probably carry oil from the point of origin and affect seals and bears far from the original problem area. A second possible source of large-scale oil pollution is spills from ice-breaking tankers if year-round shipping of oil becomes a reality. Smaller spills might originate from support vessels working around offshore rigs or keeping shipping routes open. Because lead systems and polynyas probably offer greater ease of passage when the sea is ice covered, it seems likely that ships moving oil will travel through them during the winter.

## Habitat Modification

In recent years, ingeniously engineered structures have been developed to facilitate offshore drilling for oil through the winter when the sea is covered with ice. The best known is probably the artificial island, made of gravel deposited at the drill site. Of greater interest, in terms of the future, is the mobile arctic caisson known as the *Molikpaq* (p. 178). This structure was designed for drilling to depths of up to 6,000 meters in 20 to 40 meters of water and appears to represent the direction in which this technology is headed. Although polar bears are seen in the vicinity of the *Molikpaq* throughout the winter, most do not remain in the area for long. There is one significant exception, however. By late winter, during periods of cold calm weather, most of the cracks in the sea ice that polar bears hunt beside freeze over. During these periods, the flow of annual ice from east to west past semipermanent sites such as the artificial islands or the *Molikpaq* results in local habitat modification. Cracks form on the

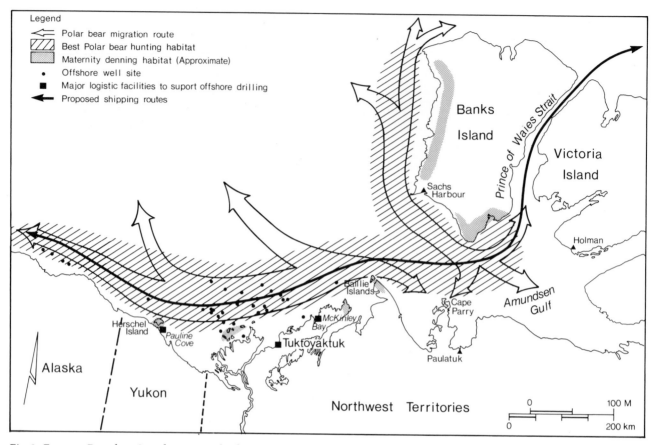

Legend

⟵▱ Polar bear migration route
▨ Best Polar bear hunting habitat
▨ Maternity denning habitat (Approximate)
• Offshore well site
■ Major logistic facilities to suport offshore drilling
⬅ Proposed shipping routes

Banks Island

Prince of Wales Strait

Victoria Island

Sachs Harbour

Holman

Amundsen Gulf

Baillie Islands

Cape Parry

McKinley Bay

Paulatuk

Herschel Island

Pauline Cove

Tuktoyaktuk

Alaska

Yukon

Northwest Territories

0          100 M
0          200 km

Fig 8. Eastern Beaufort Sea showing polar bear migration routes

downstream side of the structure and sometimes these can be quite sizable. There is always at least some open water near the structure because of the constant passage of ice. The cracks close to the rig are used routinely by small numbers of seals, probably because of the easy access to air. Sometimes up to thirty or forty seals have been present for a day or two at a time. Probably because of the availability of seals, polar bears are also attracted and they often hunt right at the base of the rig, where the open crack of water is. When the ice in surrounding areas remains frozen, bears may remain by the *Molikpaq* to hunt for days or even weeks.

As one of the conditions of receiving a drilling permit, a contractor is required to maintain a regular record of sightings of wildlife. The field staff are so interested in polar bears, or afraid of them,

that bear sightings are recorded with a fair degree of accuracy. For example, between 26 October 1984 and 30 April 1985 the *Molikpaq* crew reported 176 polar bears. They sighted bears of all ages and sex classes except adult females with cubs of the year. Twenty-eight were thought to be hunting and 17 were feeding on seals they had caught. It was a photographer's dream. The working platform was safe, several meters above the ice. The workers took all kinds of pictures of bears hunting and eating seals, cubs playing, and even an entertaining video of foxes scavenging from the seal kills and being run off by the bears. One adult female with a radio collar, and accompanied by a yearling cub, was observed on six different days between 25 December and 2 January. Another female and cub, possibly the same ones, were seen on six different days between 21 February and 10 April. As spring advanced and more new cracks appeared in the ice, the concentrations of bears drifted away from the *Molikpaq* again.

Polar bears are quick to learn new hunting tricks and clearly it did not take them very long to realize that neither the *Molikpaq* nor even the artificial islands were very dangerous. More important, they also learned there was good seal hunting habitat by the rigs. This has two quite important possible consequences. One is that some bears may become used to the presence of humans and may lose their fear of them. Successful hunting around permanent offshore drilling rigs will probably result in some polar bears being attracted to any new drill sites they encounter. If bears are attracted, it could be more dangerous for men to work on the ice in the area. Bears might be shot if they threatened people. Another risk to the bears of being attracted is that if there was even a fairly small blowout or spill, they could easily come into contact with it. In this circumstance, even small, perhaps undetectable quantities of oil could be quite dangerous to individual polar bears.

## Toxic Chemicals

Almost thirty years ago, the insecticide dichloro-diphenyltrichlorethane, better known as DDT, was found in the tissues of Adelie penguins in Antarctica. This finding signaled the universal transport of toxic chemicals through the oceans of the world. Studies of contaminants in polar bear tissues have turned up some disturbing evidence of other environmentally dangerous chemicals.

Animals at each ascending level of the food pyramid have higher levels of toxic chemicals in their tissues than those below them. They concentrate the contaminants present in all the animals and plants they eat. Consequently,

predators at the top of the food chain develop the highest concentration of all. By periodically checking the chemical concentrations in these predators, we can directly monitor the chemical health or deterioration of the ecosystem.

Tissue samples collected from polar bears off the west and north coasts of Alaska between 1967 and 1972 were tested by Jack Lentfer for mercury contamination. He found that levels were significantly higher in bears living along the north coast of the Beaufort Sea than in those from the Chukchi Sea. Shortly afterward, Richard Eaton, a physician from Edmonton, examined mercury levels in samples of polar bear hair collected from several parts of the Arctic. He compared the results with specimens taken from museum samples collected between 1910 and 1927. Like Lentfer, he found that mercury concentrations in polar bear hair were significantly higher in the Beaufort Sea than in other areas sampled from the eastern Canadian Arctic. However, the concentrations found in the recent specimens were not significantly different from museum specimens collected in the same general areas fifty or more years previously. He concluded that the different levels of mercury contamination were attributable to variations in natural geologic sources rather than human origins. The possible effects on polar bears of high concentrations of mercury are unknown. Mercury levels in ringed seal livers remain a concern for human health, but we don't know what the longer term effects on polar bears might be.

Gerald Bowes and Chuck Jonkel examined tissues collected from polar bears, seals, and arctic char from several areas of the Canadian Arctic from 1968 to 1972. Polychlorinated biphenyls (PCBs) and DDT were present in all the samples, although the concentrations of PCBs were higher. Concentrations of PCBs were much higher in polar bears than in seals or arctic char. The highest levels of PCBs were from cubs and nursing young bears due to concentration of the contaminant in the milk. After the cubs were weaned, the levels declined for a period. They increased again in older bears as they ate more seals and accumulated more chemicals through time. In general, the residue levels were higher in bears from the eastern Canadian Arctic, and subarctic areas such as Hudson Bay, than they were elsewhere.

Ray Schweinsburg and Ross Norstrom, a toxic chemical specialist with the Canadian Wildlife Service, collected a second sample of tissues from polar bears in the Canadian Arctic in the 1980s. They wanted to see if the levels of contaminants had changed over the intervening fifteen-year period since the first study was begun. They found that the levels of most contaminants were higher than in the previous sampling period. The levels of one class of chlorinated hydrocarbons, the chlordanes, had doubled in Hudson Bay and Baffin Bay. These concentrations are now high enough that there are concerns about possible detrimental effects on both the polar bears and the local Inuit, who also consume seals. Clearly, there is a long-term concern for the health of the whole arctic marine ecosystem as it is threatened by increasing concentrations of harmful chemicals.

## Some Unanswered Questions

To survive in an extreme environment like the Arctic, a predator must investigate anything novel or different in its environment, for it never knows when it may discover a new opportunity. Thus the polar bear is curious about everything it encounters. It is also a scavenger that will eat an extraordinary variety of unnatural foods. Some of these are so obviously unhealthy that it is unclear why the bear eats them. For example, in the stomachs or scats of polar bears we have found things like styrofoam, pieces of plastic, and even parts of an old car battery, which killed the animal. Another fairly regular observation from Inuk hunters that quite puzzles me is that the bears will often bite and chew cans of snowmobile oil.

While sometimes humorous, such anecdotes raise a number of serious concerns. For example: Are there smells in unrefined crude oil that might attract a bear from some distance to investigate a spill or blowout? Would a bear eat a dead bird or seal that was covered in oil? Would a bear avoid a lead that was covered with oil or would it swim through it and foul its fur? At present, there are no answers to these critical questions.

# The Future

When the First International Meeting on the Conservation of the Polar Bear was held in 1965, there was serious concern that this magnificent animal might be endangered, mainly because of excessive hunting. Now, just over twenty years later, as a result of the tremendous international effort in research, management, and the establishment of the International Agreement on the Conservation of Polar Bears and Their Habitat, the species seems reasonably secure. Yet in spite of the impressive progress that has been made to date, there are ongoing developments and future concerns to be noted and worked on.

## Management of Hunting

Polar bears remain completely protected in the Soviet Union and it seems unlikely that this will change. In Svalbard, the moratorium on hunting polar bears continues. At some future time, Norway may allow the hunting of polar bears on Svalbard again. If this were to happen, hunting would be strictly controlled and governed by the results of scientific studies of population size and dynamics. There would be a controversy if hunting did begin again, but it is highly unlikely that the bear population would become endangered because of hunting.

In the meantime, polar bear populations in the Soviet Union and Svalbard have recovered to the point where bear-human conflicts have become a significant concern. Some management measures will likely have to be taken, but these will be of local importance only. The need to develop a plan for problem bears underscores the fact that simply stopping hunting does not solve all the problems when one is dealing with a large carnivore.

Hunting of polar bears by Inuit in Greenland has continued at a fairly constant level for many years. Although the population data are generally poor, the harvest seems to be acceptable so far. Recovery by Greenland hunters of ear tags from polar bears tagged in Canada indicates the need for a sharper focus on the assessment of population size, discreteness, and movements in the region of Baffin Bay and Davis Strait. This is especially important in Northwest Greenland, where it appears that Inuit from Canada and Greenland are hunting bears from the same population.

In Canada and Alaska, where the greatest numbers of polar bears have been killed in recent years, there have also been a number of developments that bode well for the conservation of polar bears. Inuit hunters themselves are becoming much more directly involved with the management of various species of wildlife, including polar bears. As a result, a whole new spirit of cooperation has developed between the traditional and scientific views of wildlife management. This has already had several positive results. For example, when population studies on the eastern coast of Baffin Island showed that polar bears were being overharvested, the Hunters and Trappers Associations of Broughton Island and Clyde River voted to reduce their quotas by two-thirds for seven years to allow the population to recover. At that time, they want the population reassessed so they can decide how to proceed.

In the Beaufort Sea, the polar bear population is shared internationally between the United States (Alaska) and Canada. The bears are hunted by both the Inupiat of Alaska and the Inuvialuit of Canada. In Canada, there are strict management regulations governing quotas, closed seasons, and protection of females with cubs and bears in dens. In Alaska, since the passage of the Marine Mammal Protection Act, native people can kill polar bears for subsistence purposes without restriction. Thus a potential for overharvest exists—and it would all be quite legal.

It could take decades to negotiate an international agreement between the United States and Canada and it might not even be possible without changes in legislation. Furthermore, such an intergovernmental pact could have a negative aspect from the point of view of the Inuit. They might see it as another edict from government rather than as coming from the people. The Inuit hunters of both areas remain concerned about the matter. They depend on the bears as a source of income, and they know how much negative publicity would be aimed at them if the population became depleted as a result of overhunting. In response, the Alaskan and Canadian Inuit have recently negotiated an unofficial agreement between themselves with which to safely manage the polar bears of the Beaufort Sea. It was formally signed in Inuvik, Northwest Territories, on 29 January 1988. They used all the scientific information available, and heeded the advice of the scientists. But they took the leadership role to conserve the bears. Clearly, this is a landmark development in polar bear conservation as well as in the direct involvement of native people in wildlife management in the Arctic. It will likely provide a model for the solution of similar issues involving polar bears elsewhere, as well as other species of arctic wildlife (see the Appendix for the entire text).

In Canada, Inuit hunters can guide nonresident sport hunters on polar bear hunts. These bears are taken as part of the annual quota. This economic opportunity has greatly increased awareness of the tremendous importance of managing the hunted populations properly. The Inuit hunters know their business will disappear if there is concern about the safety of the population. The guided sport hunt does not increase the number of bears taken; the tags issued come out of the quota that has already been established for each village. The economic benefit to the Inuit is significant, however. Instead of $1,000 or so for a hide, the Inuit are paid $16,000 or more for the hunt. Guides, helpers, and many others in the community benefit. In a cash-poor economy with few opportunities for diversification, the income from guiding hunters for several species is extremely important.

The use of mechanized vehicles such as snowmobiles is not allowed. The hunt has to be conducted with a dog team. The opportunity to spend up to two weeks traveling in that fashion on the sea ice and get to know a bit about the life of an Inuk hunter is a rare one. To me, such a trip would be more rewarding than taking a polar bear hide home for a trophy.

There is another irony that opponents of guided sport hunts should remember. Each village has an allotment of polar bear tags assigned to it and the hunters have all winter to get them, so it is unusual not to fill the quota. When tags are assigned to sport hunts, they are not returned to the Inuit hunters if the sport hunters are not successful. Each year several are not. What this means is that through the sport-hunting program, more benefits accrue to the Inuit while fewer polar bears are killed.

## Environmental Concerns

Despite the successes that have been achieved in management, there are still a few areas that warrant serious continued attention. Polar bear populations will remain vulnerable because of their slow reproductive rate. Widespread environmental disruption, or overhunting, could quickly cause local populations to plummet. Because of the slow reproductive rate of females, recovery of a population would take several years. If other critical components of the ecosystem were seriously disrupted, the recovery time could be even longer. For this reason, further consideration needs to be given to increasing the level of protection accorded to a number of critical areas for polar bears and the marine ecosystems upon which they depend. Norway, the Soviet Union, and Denmark have taken important steps in this area by protecting the vital maternity denning areas on Køngsøya (Svalbard), Wrangel and Herald islands (USSR), and Melville Bay (Greenland). Increased protection is still needed in Canada for the large denning area near Cape Churchill, Manitoba, and possibly some other smaller denning areas as well.

Some kind of protection, or management plan, for polar bears and other marine mammals and birds, needs to be developed for the critical feeding areas. For the most part, these tend to be found along or adjacent to the circumpolar system of shore leads and polynyas that Uspenski calls the "Arctic ring of life." In some places, the leads and polynyas overlie areas of intense interest for offshore hydrocarbon production. They are also likely to be used extensively as shipping routes during the winter. Whether these activities will be detrimental to polar bears or other species is unknown. Although the International Agreement on the Conservation of Polar Bears obligates countries to "protect the ecosystems of which polar bears are a part," it is not clear just how this is to be done.

Global pollution remains the most insidious threat on the horizon, however, not just for polar bears but for man as well. Continually rising levels of toxic chemicals in the oceans of the world represent a time bomb for many species, but especially those that, like the polar bear, are at or near the top of the food chain. The problem cannot be attacked in the Arctic. It must be dealt with in the industrialized areas of the world.

## The New Polar Bear Watchers

Undoubtedly, the most the most important factor helping to ensure the permanent welfare of the polar bear is its enormous popularity. People who know they will never see a live polar bear are as interested in their conservation as those who have. The public as a whole has an insatiable desire to see and read about polar bears. Every year, film crews and writers come to Churchill, Manitoba, to do the same stories as their predecessors and they all seem to be well received. In our lab, we always know when it is fall because the geese start to fly south, the leaves turn color, and the photographers begin to phone for information about polar bears at Churchill.

Public interest in wild polar bears, and the ease of seeing them in places like Churchill, has set the scene for what may be one of the most significant developments in polar bear conservation. The population of the town of Churchill has been slowly declining for several years and there are not a lot of directions into which new businesses can diversify. However, Churchill does have polar bears. Each fall, several thousand people come, both individually and on guided tours, to see them. The motels are busy all summer with people who come to see the white whales and birds. Occupancy rates remain high into early November as polar bear watchers arrive. This gives local motels, restaurants, and shops a longer season than that enjoyed by most tourist destinations in southern Canada.

Not too many years ago, polar bears were regarded as pests. Now they are seen as the most valuable attraction that Churchill has going for it. The town motto is, "Churchill, Polar Bear Capital of the World." Tour buses leave town every day through the fall to drive around the local network of roads in search of bears. In one of the most imaginative developments, Len Smith has ingeniously reconstructed the remains of a number of discarded machines of various sorts into a series of Tundra Buggies, which ride high off the ground on six-foot-tall agricultural tires. Every day, these vehicles use designated routes to travel far out along the coast from Churchill, sometimes camping for a week or so. They provide some of the most spectacular wildlife viewing in the world. The tours are famous now and are booked over a year in advance.

There are several other areas of the circumpolar Arctic where similar developments are possible, if they are sensitive to the needs of the animals and the surrounding environment. No doubt some will be followed up on in time, thereby creating an additional incentive to ensure the development of management plans for polar bears that include the interests of all the groups concerned with each population.

The Canadian Polar Bear Management Zone A1 runs from about the Manitoba-Ontario border in the south to Rankin Inlet in the north (see Fig. 7). This means that the population is shared between Manitoba and the Northwest Territories and that there are two major uses of polar bears. First, there is a quota of sixty-two polar bears in the Inuit settlements of that zone. Most of the bears taken are fairly young and are hunted in the late fall. Thus many of the hides are not large or in as good condition as those taken in the winter, so the overall value is reduced. It is unlikely their total value would reach $60,000. Polar bear hunting is also very important culturally, although it is difficult to put a dollar value on such an activity. The second use of polar bears is as a tourist attraction. The tourism business in Churchill is likely worth $1 million or more each year. What this means is that the governments of the Northwest Territories and the Province of Manitoba, which are jointly responsible for this population of polar bears, will continue to be pressured from both inside and outside their jurisdictions, to ensure the population is properly conserved. Like the large mammals of Africa, polar bears have developed a large and international constituency. This, coupled with their significant economic value as a tourist attraction, is probably the best insurance of all for the maintenance of healthy polar bear populations in the future.

## A Personal Epilogue

The story of the polar bear through recent years is a rewarding one and an example to be emulated in modern conservation of large carnivores. Starting from little knowledge and a deep concern about the survival of the species, scientists and managers have made progress by virtue of the understanding gained from basic research. Management changes in most areas have considered several legitimate, if not always totally compatible, interests in polar bears. As we continue to acquire a greater understanding of polar bears from our studies in the future, and learn more about the merits and weaknesses of our present management plans, they too will evolve. I also harbor the deep hope that the results of the research on polar bears over the last twenty years or so will help us in another critical area: understanding the importance of protecting tracts of habitat large enough to ensure the survival of other species, of large carnivores that compete with man for space. As the burgeoning human population continues to overwhelm the remaining wilderness areas left on the planet, we have limited time left in which to act. Perhaps there is still time to follow the example of the great white bear. If so, we must learn to be an integral part of the environment instead of fighting it, and leave it unspoiled behind us, like a wind-blown set of polar bear tracks across the the drifting ice floes of the Arctic.

# Appendix

## The International Agreement on the Conservation of Polar Bears and Their Habitat

THE GOVERNMENTS of Canada, Denmark, Norway, the Union of Soviet Socialist Republics, and the United States of America,

RECOGNIZING the special responsibilities and special interests of the States of the Arctic Region in relation to the protection of the fauna and flora of the Arctic Region;

RECOGNIZING that the polar bear is a significant resource of the Arctic Region which requires additional protection;

HAVING DECIDED that such protection should be achieved through coordinated nation measures taken by the States of the Arctic Region;

DESIRING to take immediate action to bring further conservation and management measures into effect;

HAVE AGREED AS FOLLOWS:

### ARTICLE I

1. The taking of polar bears shall be prohibited except as provided in Article III.
2. For the purpose of this Agreement, the term "taking" includes hunting, killing and capturing.

### ARTICLE II

Each Contracting Party shall take appropriate action to protect the ecosystems of which polar bears are a part, with special attention to habitat components such as denning and feeding sites and migration patterns, and shall manage polar bear populations in accordance with sound conservation practices based on the best available scientific data.

### ARTICLE III

1. Subject to the provisions of Articles II and IV, any Contracting Party may allow the taking of polar bears when such taking is carried out:

   (a) for *bona fide* scientific purposes; or

   (b) by that Party for conservation purposes; or

   (c) to prevent serious disturbance of the management of other living resources, subject to forfeiture to that Party of the skins and other items of value resulting from such taking; or

   (d) by local people using traditional methods in the exercise of their traditional rights and in accordance with the laws of that Party; or

   (e) wherever polar bears have or might have been subject to taking by traditional means by its nationals.

2. The skins and other items of value resulting from taking under subparagraphs (b) and (c) of paragraph 1 of this Article shall not be available for commercial purposes.

## ARTICLE IV

The use of aircraft and large motorized vessels for the purpose of taking polar bears shall be prohibited, except where the application of such prohibition would be inconsistent with domestic laws.

## ARTICLE V

A Contracting Party shall prohibit the exportation from, the importation and delivery into, and traffic within, its territory of polar bears or any part or product thereof taken in violation of this Agreement.

## ARTICLE VI

1. Each contracting Party shall enact and enforce such legislation and other measures as may be necessary for the purpose of giving effect to this Agreement.

2. Nothing in this Agreement shall prevent a Contracting Party from maintaining or amending existing legislation or other measures or establishing new measures on the taking of polar bears so as to provide more stringent controls than those required under the provisions of this Agreement.

## ARTICLE VII

The Contracting Parties shall conduct national research programmes on polar bears, particularly research relating to the conservation and management of the species. They shall as appropriate coordinate such research with the research carried out by other Parties, consult with other Parties on the management of migrating polar bear populations, and exchange information on research and management programmes, research results and data on bears taken.

## ARTICLE VIII

Each Contracting Party shall take actions as appropriate to promote compliance with the provisions of this Agreement by nationals of States not party to this Agreement.

## ARTICLE IX

The Contracting Parties shall continue to consult with one another with the object of giving further protection to polar bears.

## ARTICLE X

1. This Agreement shall be open for signature at Oslo by the Governments of Canada, Denmark, Norway, the Union of Soviet Socialist Republics and the United States of America until 31st March 1974.

2. This Agreement shall be subject to ratification or approval by the signatory Governments. Instruments of ratification or approval shall be deposited with the Government of Norway as soon as possible.

3. This Agreement shall be open for accession by the Governments referred to in paragraph 1 of this Article. Instruments of accession shall be deposited with the Depository Government.

4. This Agreement shall enter into force ninety days after the deposit of the third instrument of ratification, approval or accession. Thereafter, it shall enter into force for a signatory or acceding Government on the date of deposit of its instrument of ratification, approval or accession.

5. This Agreement shall remain in force initially for a period of five years from its date of entry into force, and unless any Contracting Party during that period requests the termination of the Agreement at the end of that period, it shall continue in force thereafter.

6. On the request addressed to the Depository Government by any of the Governments referred to in paragraph 1 of this Article, consultations shall be conducted with a view to convening a meeting of representatives of the five Governments to consider the revision or amendment of this Agreement.

7. Any Party may denounce this Agreement by written notification to the Depository Gov-

ernment at any time after five years from the date of entry into force of this Agreement. The denunciation shall take effect twelve months after the Depository Government has received this notification.

8. The Depository Government shall notify the Governments referred to in paragraph 1 of this Article of the deposit of instruments of ratification, approval or accession, for the entry into force of this Agreement and of the receipt of notifications of denunciation and any other communications from a Contracting Party specially provided for in this Agreement.

9. The original of this Agreement shall be deposited with the Government of Norway which shall deliver certified copies thereof to each of the Governments referred to in paragraph 1 of this Article.

10. The Depository Government shall transmit certified copies of this Agreement to the Secretary-General of the United Nations for registration and publication in accordance with Article 102 of the Charter of the United Nations.

[The Agreement came into effect in May 1976, three months after the third nation required to ratify did so in February 1976. All five nations ratified by 1978. After the initial period of five years, all five Contracting Parties met in Oslo, Norway, in January 1981, and unanimously reaffirmed the continuation of the Agreement.]

# Management Agreement for Polar Bears in the Southern Beaufort Sea

Section II

Agreement Between:

The Inuvialuit Game Council
of Inuvik, NWT, Canada

and

The North Slope Borough Fish and Game
Management Committee of Barrow, Alaska, USA

The Inuvialuit of Canada and the Inupiat of the United States

NOTING that both groups have traditionally harvested a portion of polar bears from the same population in the southern Beaufort Sea; and,

NOTING that the continued hunting of polar bears is essential to maintain the dietary, cultural and economic base of the groups; and,

NOTING that the maintenance of a sustained harvest for traditional users in perpetuity requires that the number of polar bears taken annually not exceed the productivity of the population; and,

NOTING that the International Agreement on the Conservation of Polar Bears makes provision for cooperation in the research and management of shared populations; and,

NOTING that nothing in this Agreement shall be read to abrogate the responsibilities of Federal, Provincial or State authorities under existing or future statutes; and,

NOTING that the Inuvialuit and the Inupiat will have a long-term fundamental influence on the maintenance and use of this resource and that the efforts of other parties will also be required to ensure effective conservation;

HAVE AGREED AS FOLLOWS:

## ARTICLE I

*Definitions*

a) The species considered in this Agreement is the polar bear (*Ursus maritimus*).

b) The area covered by this Agreement is the southern Beaufort Sea from approximately Baillie Islands, Canada, in the east to Icy Cape, USA in the west.

c) The people covered by this Agreement are the Inuvialuit of Canada and the Inupiat of the North Slope of Alaska.

d) The settlements whose hunting practices may be affected by this Agreement are Barrow, Nuiqsut, Wainwright, Atqasuk and Kaktovik in the United States and Inuvik, Aklavik, Tuktoyaktuk and Paulatuk in Canada.

e) Sustained yield is a level of taking which does not exceed recruitment and is consistent with population ranges determined to be optimal and sustainable.

f) The Joint Commission shall consist of two (2) representatives designated by each of the Inuvialuit Game Council and the North Slope Borough Fish and Game Management Committee. The Technical Advisory Committee shall be appointed by the Joint Commission.

## ARTICLE II

*Objectives*

a) To maintain a healthy viable population of polar bears in the southern Beaufort Sea in perpetuity.

b) To provide the maximum amount of protection to female polar bears.

c) To minimize detrimental effects of human activities, especially industrial activities, on important polar bear habitat.

d) To manage polar bears on a sustained yield basis in accordance with all the best information available.

e) To encourage the collection of adequate technical information on a timely basis to facilitate management decisions.

f) To further refine the eastern and western boundaries of the population of polar bears.

g) To encourage the wise use of polar bear products and by-products within the context of management on a sustained yield basis.

h) To facilitate the exchange of polar bear meat and products between traditional users in

Alaska and Canada (Enabling legislation required).

i) To legalize the sale of polar bear hides and by-products by the traditional Alaskan users in Alaska (Enabling legislation required).

j) To facilitate the export of polar bear hides and other polar bear products from the Western Arctic of Canada into the USA (Enabling legislation required).

k) To consider at a later date a limited legalized Alaskan sport harvest of polar bears which emphasizes benefits to local hunters of the area (Enabling legislation required for Federal management).

## ARTICLE III

*Regulations*

To conserve this population of polar bears, the Inuvialuit and the Inupiat have agreed as follows:

a) All bears in dens or constructing dens are protected.

b) Family groups made up of females and cubs-of-the-year or yearlings are protected. The birthdate of cubs is fixed at January 1 and cubs less than five feet (152 cm.) in straight line body length are protected.

c) The hunting season shall extend from December 1 to May 31 in Canada and from September 1 to May 31 in Alaska.

d) The annual sustainable harvest shall be determined by the Technical Advisory Committee in consultation with the Joint Commission and shall be divided between Canada and Alaska according to annual review of scientific evidence. Allocation agreements shall be negotiated and ratified prior to September 1 annually. Each signatory to this Agreement shall determine for itself the distribution of the harvest within its jurisdiction.

e) These regulations do not preclude either party from unilaterally introducing additional conservation practices within their own jurisdictions.

f) Any readjustment of the boundaries pursuant to the above may necessitate a readjustment of user allocations under the management plan.

g) The use of aircraft or large motorized vessels for the purpose of taking polar bears shall be prohibited.

h) Each jurisdiction shall prohibit the exportation from, the importation and delivery into, and traffic within, its territory of polar bears or any part or product thereof taken in violation of this Agreement.

i) Polar bears in villages during closed seasons should be deterred from the area.

j) Polar bears threatening human safety or property may be taken at any time of the year and may be counted against the village allocation as ascribed by the Joint Commission.

## ARTICLE IV

*Collection of Data and Sharing of Information*

a) The following data will be recorded for each bear killed: sex, date and location of kill, and hunter's name.

b) The following shall be collected from each bear killed: an undamaged post-canine tooth, ear tags or lip tattoos if the tags are missing, other specimens as agreed to by the hunters of either jurisdiction for additional studies.

c) A summary of all harvest information from each jurisdiction shall be exchanged annually.

d) The number of collars deployed for research purposes shall be limited to the minimum number necessary to provide accurate population information.

## ARTICLE V

*Duration of Agreement*

a) This Agreement shall enter into force when it has been signed by the representatives of both parties.

b) This Agreement shall remain in force unless either Contracting Party requests it be terminated.

c) Amendments to the Agreement may be proposed by either signatory and accepted or rejected by mutual agreement after consultation with the membership of the

Inuvialuit Game Council and the North Slope Borough Fish and Game Management Committee.

The Alaskan signatories of this document have no authority, to bind and do not purport to bind the North Slope Borough to any agreement which would otherwise be in violation of the exclusive federal treaty power established by the United States Constitution, but are acting solely as representatives of the local traditional user group of the polar bear resource in furthering the consultation, management, and information exchange goals of the International Agreement on the Conservation of Polar Bears.

SIGNED on this 29th day of January, 1988 in the Town of Inuvik, Northwest Territories.

On behalf of the North Slope Inupiat

NOLAN SOLOMON
Chairman
North Slope Borough
Fish and Game Management Committee.

BENJAMIN P. NAGEAK
Director
Department of Wildlife Management

On behalf of the Inuvialuit Game Council

ALEX AVIUGANA
Chairman
Inuvialuit Game Council

ANDY CARPENTER
Vice-Chairman
Wildlife Management Advisory Council (N.W.T.)

# Bibliography

Amstrup, S. C., I. Stirling, and J. W. Lentfer. 1986. Size and trends of Alaskan polar bear populations. *Wildlife Society Bulletin* 14:241–54.

Baker, B. E., C. R. Harington, and A. L. Symes. 1963. Polar bear milk. I. Gross composition and fat constitution. *Canadian Journal of Zoology* 41:1035–39.

Bannikov, A. G., A. A. Kishchinsky, and S. M. Uspenski, eds. 1969. *Belyi medvi' i ego okhrana v Sovetskoy Arktike (The Polar Bear and Its Conservation in the Arctic).* Leningrad: Gidrometeorologizdat. Unedited translation by Government of Canada Translation Bureau, No. 5776, August 1970.

Best, R. C. 1977. Ecological aspects of polar bear nutrition. In *Proceedings of the 1975 Predator Symposium,* edited by R. L. Phillips and C. Jonkel, 201–11. Missoula: Montana Forest and Conservation Experiment Station, University of Montana.

————. 1982. Thermoregulation in resting and active polar bears. *Journal of Comparative Physiology,* ser. B 146:63–73.

————. 1985. Digestibility of ringed seals by the polar bear. *Canadian Journal of Zoology* 63:1033–36.

Best, R. C., K. Ronald, and N. A. Øritsland. 1981. Physiological indices of activity and metabolism in the polar bear. *Journal of Comparative Biochemistry and Physiology,* ser. A 69:177–85.

Blix, A. S., and J. W. Lentfer. 1979. Modes of thermal protection in polar bear cubs at birth and on emergence from the den. *American Journal of Physiology* 236:67–74.

Boas, F. 1964. *The Central Eskimo.* Lincoln, Nebr.: University of Nebraska Press. Originally published in Washington, D.C., as part of the 6th Annual Report of the Bureau of Ethnology, Smithsonian Institution, 1888.

Bowes, G. W., and C. J. Jonkel. 1975. Presence and distribution of polychlorinated biphenyls (PCB) in arctic and subarctic marine food chains. *Journal of the Fisheries Research Board of Canada* 32:2111–23.

Bromley, M. 1985. *Safety in Bear Country: A Reference Manual.* Yellowknife: Government of the Northwest Territories.

Brooks, J. W. 1972. Infrared scanning for polar bears. In *Bears—Their Biology and Management,* edited by S. Herrero, 106–33. International Union for the Conservation of Nature, n.s., no. 23. Morges, Switzerland.

Cook, H. W., J. W. Lentfer, A. M. Pearson, and B. E. Baker. 1970. Polar bear milk. IV. Gross composition, fatty acid, and mineral constitution. *Canadian Journal of Zoology* 48:217–19.

Davids, R. C. 1982. *Lords of the Arctic.* New York: Macmillan Co.

Davies, J. C. 1986. An electric fence to deter polar bears. *Wildlife Society Bulletin* 14:406–9.

DeMaster, D. P., M. C. S. Kingsley, and I. Stirling. 1980. A multiple mark and recapture estimate applied to polar bears. *Canadian Journal of Zoology* 58:633–38.

DeMaster, D. P., and I. Stirling. 1981. *Ursus maritimus. Mammalian Species* 145:1–7.

Doutt, J. K. 1967. Polar bear dens on the Twin Islands, James Bay, Canada. *Journal of Mammalogy* 48:468–71.

Dunbar, M. J. 1968. *Ecological Development in Polar Regions.* Englewood Cliffs, N.J.: Prentice-Hall.

Eaton, R. D. P., and J. P. Farant. 1982. The polar bear as a biological indicator of the environmental mercury burden. *Arctic* 35:422–25.

Flyger, V., M. W. Schein, A. W. Ericksen, and T. Larsen. 1967. Capturing and handling polar bears. *Transactions of the Thirty-second North American Wildlife Conference* 32:107–19.

Folk, G. E., M. D. Brewer, and D. Sanders. 1970. Cardiac physiology of polar bears in winter dens. *Arctic* 23:130.

Folk, G. E., M. A. Folk, and J. J. Minor. 1972. Physiological condition of three species of bears in winter dens. In *Bears—Their Biology and Management,* edited by S. Herrero, 107–24. International Union for the Conservation of Nature, n.s., no. 23. Morges, Switzerland.

Frame, G. W. 1969. Occurrence of polar bears in the Chukchi and Beaufort seas, summer, 1969. *Journal of Mammalogy* 53:187–89.

Freeman, M. M. R. 1973. Polar bear predation on beluga in the Canadian Arctic. *Arctic* 26:163–64.

Furnell, D. J., and D. Oolooyuk. 1980. Polar bear predation on ringed seals in ice-free water. *Canadian Field Naturalist* 94:88–89.

Furnell, D. J., and R. E. Schweinsburg. 1984. Population dynamics of central Arctic polar bears. *Journal of Wildlife Management* 48:722–28.

Gray, A. P. 1972. Mammalian hybrids: A check-list with bibliography. Technical Communication no. 10 rev. Edinburgh: Commonwealth Bureau Animal Breeding and Genetics.

Hallowell, C. R. 1926. Bear ceremonialism in the northern hemisphere. *American Anthropologist* 28:1–175.

Hansson, R., and J. Thomassen. 1983. Behavior of polar bears with cubs in the denning area. *Proceedings of the Fifth International Conference on Bear Research and Management* 5:246–54.

Harington, C. R. 1962. A bear fable? *The Beaver*, Winter 1962, 4–7.

———. 1966. The bear behind the paw. *The Beaver*, Autumn 1966, 14–15.

———. 1968. Denning habits of the polar bear (*Ursus maritimus* Phipps). Canadian Wildlife Service Report Series, no. 5. Ottawa.

Herrero, S. 1985. *Bear Attacks: Their Causes and Avoidance*. Piscataway, N.J.: Winchester Press.

Heyland, D., and K. Hay. 1976. An attack by a polar bear on a juvenile beluga. *Arctic* 29:56–57.

Hughes, C. C. 1960. *An Eskimo Village in the Modern World*. Ithaca, N.Y.: Cornell University Press.

Hurst, R. J., M. L. Leonard, P. D. Watts, P. Beckerton, and N. A. Øritsland. 1982. Polar bear locomotion: Body temperature and energetic cost. *Canadian Journal of Zoology* 60:40–44.

Hurst, R. J., and N. A. Øritsland. 1982. Polar bear thermoregulation: Effect of oil on the insulative properties of fur. *Journal of Theoretical Biology* 7:201–8.

Hurst, R. J., N. A. Øritsland, and P. D. Watts. 1982a. Body mass, temperature and cost of walking in polar bears. *Acta Physiologica Scandinavica* 115:391–95.

———. 1982b. Metabolic and temperature responses of polar bears to crude oil. In *Land and Water Issues Related to Energy Development*, edited by P. J. Rand, 263–80. Ann Arbor: Ann Arbor Science.

IUCN Survival Service Commission. 1968–85. *Proceedings of the First to the Ninth Working Meetings of the Polar Bear Specialists Group*. Gland, Switzerland: Publications of the International Union for the Conservation of Nature.

Jenness, P., A. W. Erickson, and J. J. Craighead. 1972. Some comparative aspects of milk from four species of bears. *Journal of Mammalogy* 53:39–47.

Jonkel, C. J., G. B. Kolenosky, R. J. Robertson, and R. H. Russell. 1972. Further notes on polar bear denning habits. In *Bears—Their Biology and Manage-*

*ment*, edited by S. Herrero, 142–58. International Union for the Conservation of Nature, n.s., no. 23. Morges, Switzerland.

Jonkel, C. J., P. Smith, I. Stirling, and G. B. Kolenosky. 1976. Notes on the present status of the polar bear in James Bay and the Belcher Islands. Canadian Wildlife Service Occasional Paper no. 26.

Kiliaan, H. P. L. and I. Stirling. 1978. Observations on overwintering walruses in the eastern Canadian High Arctic. *Journal of Mammalogy* 59:197–200.

Kingsley, M. C. S. 1979. Fitting the von Bertalanffy growth equation to polar bear age-weight data. *Canadian Journal of Zoology* 57:1020–25.

Kolenosky, G. B., and J. P. Prevett. 1983. Productivity and maternity denning of polar bears in Ontario. *Proceedings of the Fifth International Conference on Bear Research and Management* 5:238–45.

Knudsen, B. 1978. Time budgets of polar bears (*Ursus maritimus*) on North Twin Island, James Bay, during summer. *Canadian Journal of Zoology* 56:1627–28.

Kurten, B. 1964. The evolution of the polar bear, *U. maritimus* Phipps. *Acta Zoologica Fennica* 108:3–30.

———. 1971. *The Cave Bear Story*. New York: Columbia University Press.

Larsen, H. 1969–70. Some examples of bear cult among the Eskimo and other northern people. *Folk* 11–12:27–42.

Larsen, T. 1971. Capturing, handling, and marking polar bears in Svalbard. *Journal of Wildlife Management* 35:27–36.

———. 1976. Polar bear den surveys in Svalbard, 1972 and 1973. In *Bears—Their Biology and Management*, edited by M. R. Pelton, J. Lentfer, and G. E. Folk, 199–208. International Union for the Conservation of Nature, n.s., no. 40. Morges, Switzerland.

———. 1985a. Polar bear denning and cub production in Svalbard, Norway. *Journal of Wildlife Management* 49:320–26.

———. 1985b. Abundance, range and population biology of the polar bear (*Ursus maritimus*) in the Svalbard Area. Ph.D. diss., University of Oslo.

Larsen, T., C. Jonkel, and C. Vibe. 1983. Satellite radio-tracking of polar bears between Svalbard and Greenland. *Proceedings of the Fifth International Conference on Bear Research and Management* 5:230–37.

Larsen, T., and B. Kjos-Hanssen. 1983. *Trichinella* sp. in polar bears from Svalbard, in relation to hide length and age. *Polar Research* 1:89–96.

Larsen, T., H. Tegelstrom, R. K. Juneja, and M. K. Taylor. 1983. Low protein variability and genetic similarity between populations of the polar bear (*Ursus maritimus*). *Polar Research* 1:97–105.

Latour, P. B. 1981a. Spatial relationships and behavior of polar bears (*Ursus maritimus* Phipps) concentrated on land during the ice-free season of Hudson Bay. *Canadian Journal of Zoology* 59:1763–74.

———. 1981b. Interactions between free-ranging, adult male polar bears (*Ursus maritimus* Phipps): A case of adult social play. *Canadian Journal of Zoology* 59:1775–83.

Lavigne, D. M., and N. A. Øritsland. Black polar bears. *Nature* 251:218–19.

Lentfer, J. W. 1968. A technique for immobilizing and marking polar bears. *Journal of Wildlife Management* 32:317–21.

———. 1972. Polar bear–sea ice relationships. In *Bears—Their Biology and Management*, edited by S. Herrero, 165–71. International Union for the Conservation of Nature, n.s., no. 23. Morges, Switzerland.

———. 1974. Discreteness of Alaskan polar bear populations. *International Game Biology* 11:323–29.

———. 1975. Polar bear denning on drifting sea ice. *Journal of Mammalogy* 56:716–18.

———. 1983. Alaskan polar bear movements from mark and recapture. *Arctic* 36:282–88.

Lentfer, J. W., R. J. Hensel, J. R. Gilbert, and F. E. Sorensen. 1980. Population characteristics of Alaskan polar bears. In *Bears—Their Biology and Management*, edited by C. J. Martinka and K. L. MacArthur, 109–16. Washington, D.C.: U.S. Government Printing Office.

Lewis, R. W., and J. W. Lentfer. 1967. The vitamin A content of polar bear liver: Range and variability. *Journal of Comparative Biochemistry and Physiology* 22:923–26.

Lønø, Ø. 1970. The polar bear in the Svalbard area. Norsk Polarinstitutt Skrifter no. 129. Oslo.

Lowry, L. F., J. J. Burns, and R. R. Nelson. 1987. Polar bear, *Ursus maritimus*, predation on belugas, *Delphinapterus leucas*, in the Bering and Chukchi seas. *Canadian Field Naturalist* 101:141–46.

Lunn, N. J. 1986. Observations of nonaggressive behavior between polar bear family groups. *Canadian Journal of Zoology* 64:2035–37.

Lunn, N. J., and I. Stirling. 1985. The significance of supplemental food to polar bears during the ice-free period of Hudson Bay. *Canadian Journal of Zoology* 63:2291–97.

Lutziuk, O. B. 1978. Contribution to the biology of the polar bear (*Ursus maritimus*) on Wrangel Island during the summer–autumn period. *Zoologischeskii Zhurnal* 57:597–603.

Lyster, S. 1985. *International Wildlife Law*. Cambridge: Grotius Publications.

McGhee, R. 1978. Inuit Prehistory. Scarborough, Ont.: Van Nostrand Reinhold.

Manning, D. P., J. E. Cooper, I. Stirling, C. M. Jones, M. Bruce, and P. C. McCausland. 1985. Studies on the footpads of the polar bear (*Ursus maritimus*) and their possible relevance to accident prevention. *Journal of Hand Surgery*, ser. B 10:303–7.

Manning, T. H. 1971. Geographical variation in the polar bear, *Ursus maritimus* Phipps. Canadian Wildlife Service Report Series, no. 13. Ottawa.

Øritsland, N. A. 1969. Deep body temperature of swimming and walking polar bear cubs. *Journal of Mammalogy* 50:380–82.

———. 1970. Temperature regulation of the polar bear (*Thalarctos maritimus*). *Journal of Comparative Biochemistry and Physiology*, ser. A 37:225–33.

Øritsland, N. A., F. R. Engelhardt, F. A. Juck, R. J. Hurst, and P. D. Watts. 1981. *Effect of Crude Oil on Polar Bears*. Environmental Studies, no. 24. Ottawa: Department of Indian and Northern Affairs.

Øritsland, N. A., C. Jonkel, and K. Ronald. 1976. A respiratory chamber for exercising polar bears. *Norwegian Journal of Zoology* 24:65–67.

Øritsland, N. A., and D. M. Lavigne. 1976. Radiative surface temperatures of exercising polar bears. *Journal of Comparative Biochemistry and Physiology*, ser. A 53:327–30.

Øritsland, N. A., J. W. Lentfer, and K. Ronald. 1974. Radiative surface temperatures of the polar bear. *Journal of Mammalogy* 55:459–61.

Øritsland, N. A., P. K. Stallman, and C. J. Jonkel. 1977. Polar bears: Heart activity during rest and exercise. *Journal of Comparative Biochemistry and Physiology*, ser. A 57:139–41.

Prevett, J. P., and G. B. Kolenosky. 1982. The status of polar bears in Ontario. *Naturaliste canadien (Review of Ecological Systems)* 109:933–39.

Ramsay, M. A., and D. S. Andriashek. 1986. Long distance route orientation of female polar bears (*Ursus maritimus*) in spring. *Journal of Zoology* (London), ser. A 208:63–72.

Ramsay, M. A., and R. L. Dunbrack. 1986. Physiological constraints on life history phenomena: The example of small bear cubs at birth. *American Naturalist* 127:735–43.

Ramsay, M. A., and I. Stirling. 1982. Reproductive biology and ecology of female polar bears in western Hudson Bay. *Naturaliste canadien (Review of Ecological Systems)* 109:941–46.

———. 1984. Interactions of wolves and polar bears in NE Manitoba. *Journal of Mammalogy* 65:693–94.

———. 1986. On the mating system of polar bears. *Canadian Journal of Zoology* 64:2142–51.

———. 1988. Reproductive biology and ecology of female polar bears (*Ursus maritimus*). *Journal of Zoology* (London), ser. A 214:601–34.

Randa, V. 1986. *L'Ours polaire et les Inuit*. Paris: Société d'Etudes Linguistiques et Anthropologiques de France.

Rasmussen, K. 1921. *Eskimo Folk Tales*. Copenhagen: Gylendal.

———. 1931. *The Netsilik Eskimos: Social Life and Spiritual Culture*. Report of the Fifth Thule Expedition, vol. 8. Copenhagen.

Russell, R. H. 1975. The food habits of polar bears of James Bay and southwest Hudson Bay in summer and autumn. *Arctic* 28:117–29.

Schweinsburg, R. E. 1979. Summer snow dens used by polar bears in the Canadian High Arctic. *Arctic* 32:165–69.

———. 1981. A brief history of polar bear management in the NWT. Northwest Territories Wildlife Notes, no. 2. Yellowknife.

Schweinsburg, R. E., L. J. Lee, and P. B. Latour. 1982. Distribution, movement and abundance of polar bears in Lancaster Sound, Northwest Territories. *Arctic* 35:159–69.

Shepard, P., and B. Sanders. 1985. *The Sacred Paw: The Bear in Nature, Myth, and Literature.* New York: Viking Penguin.

Smith, T. G. 1980. Polar bear predation of ringed and bearded seals in the land-fast sea ice habitat. *Canadian Journal of Zoology* 58:2201–9.

———. 1985. Polar bears, *Ursus maritimus,* as predators of belugas, *Delphinapterus leucas. Canadian Field Naturalist* 99:71–75.

———. 1987. The ringed seal, *Phoca hispida,* of the Canadian Western Arctic. *Canadian Bulletin of Fisheries and Aquatic Science,* no. 216.

Smith, T. G., and I. Stirling. 1978. Variation in the density of ringed seal (*Phoca hispida*) birth lairs in the Amundsen Gulf, Northwest Territories. *Canadian Journal of Zoology* 56:1066–71.

Soby, R. M. 1969–70. The Eskimo animal cult. *Folk* 11–12:43–78.

Stefannson, V. 1913. *My Life with the Eskimos.* New York: Macmillan Co.

———. 1921. *The Friendly Arctic.* New York: Macmillan Co.

Stirling, I. 1974a. Midsummer observation on the behavior of wild polar bears (*Ursus maritimus*). *Canadian Journal of Zoology* 52:1191–98.

———. 1974b. Polar bear research in the Beaufort Sea. In *The Coast and Shelf of the Beaufort Sea,* edited by J. C. Reed and J. E. Sater, 721–23, Arlington, Va.: Arctic Institute of North America.

———. 1977. Adaptations of Weddell and ringed seals to exploit the polar fast ice habitat in the absence or presence of surface predators. In *Adaptations within Antarctic Ecosystems,* edited by G. A. Llano, pp. 741–48. Houston: Gulf Publishing.

———. 1980. The biological importance of polynyas in the Canadian Arctic. *Arctic* 33:303–15.

———. 1984. A group threat display given by walruses to a polar bear. *Journal of Mammalogy* 65:352–53.

———. 1986. Research and management of polar bears (*Ursus maritimus*). *Polar Record* 23:167–76.

———. 1988. Attraction of polar bears to offshore drilling sites in the eastern Beaufort Sea. *Polar Record* 24:1–8.

Stirling, I., D. Andriashek, P. Latour, and W. Calvert. 1975. *Distribution and Abundance of Polar Bears in the Eastern Beaufort Sea: A Final Report to the Beaufort Sea Project.* Victoria, B.C.: Fisheries and Marine Service, Department of the Environment.

Stirling, I., and R. Archibald. 1977. Aspects of predation of seals by polar bears. *Journal of the Fisheries Research Board of Canada* 34:1126–29.

Stirling, I., and W. Calvert. 1983. Environmental threats to marine mammals in the Canadian Arctic. *Polar Record* 21:433–49.

Stirling, I., W. Calvert, and D. Andriashek. 1980. Population ecology studies of the polar bear in the area of southeastern Baffin Island. Canadian Wildlife Service Occasional Paper no. 44. Ottawa.

———. 1984. Polar bear (*Ursus maritimus*) ecology and environmental considerations in the Canadian High Arctic. In *Northern Ecology and Resource Management,* edited by R. Olsen, R. Hastings, and F. Geddes, 201–22. Edmonton: University of Alberta Press.

Stirling, I., C. Jonkel, P. Smith, R. Robertson, and D. Cross. 1977. The ecology of the polar bear (*Ursus maritimus*) along the western coast of Hudson Bay. Canadian Wildlife Service Occasional Paper no. 33. Ottawa.

Stirling, I., and H. P. L. Kiliaan. 1980. Population ecology studies of the polar bear in Northern Labrador. Canadian Wildlife Service Occasional Paper no. 42. Ottawa.

Stirling, I., M. C. S. Kingsley, and W. Calvert. 1982. The distribution and abundance of seals in the eastern Beaufort Sea, 1974–79. Canadian Wildlife Service Occasional Paper no. 44. Ottawa.

Stirling, I., and P. B. Latour. 1978. Comparative hunting abilities of polar bear cubs of different ages. *Canadian Journal of Zoology* 56:1768–72.

Stirling, I., and E. H. McEwan. 1975. The caloric value of whole ringed seals (*Phoca hispida*) in relation to polar bear (*Ursus maritimus*) ecology and hunting behavior. *Canadian Journal of Zoology* 53:1021–27.

Stirling, I., A. M. Pearson, and F. L. Bunnell. 1976. Population ecology studies of polar and grizzly bears in northern Canada. *Transactions of the Forty-first North American Wildlife Conference* 41:421–30.

Taylor, M. K., D. P. DeMaster, F. L. Bunnell, and R. E. Schweinsburg. 1987. Modelling the sustainable harvest of female polar bears. *Journal of Wildlife Management* 51:811–20.

Taylor, M. K., T. Larsen, and R. E. Schweinsburg. 1985. Observations of intraspecific aggression and cannibalism in polar bears (*Ursus maritimus*). *Arctic* 38:303–9.

Uspenski, S. M., and A. A. Kistchinski. 1972. New data on the winter ecology of the polar bear (*Ursus maritimus* Phipps) on Wrangel Island. In *Bears—Their Biology and Management,* edited by S. Herrero, 181–97. International Union for the Conservation of Nature, n.s., no. 23. Morges, Switzerland.

Uspenskii, S. M. 1977. *Belyi Medved (The Polar Bear).* Moscow: Navka. Unedited translation by Government of Canada Translation Bureau, No. 1541321, June 1978.

Van de Velde, F. 1957. Nanuk, king of the arctic beasts. *Eskimo* 45:4–15.

———. 1971. Bear stories. *Eskimo,* n.s. 1:7–11.

Vibe, C. 1967. *Arctic Animals in Relation to Climatic Fluctuations.* Meddelelser om Gronland Bind. no. 5. Copenhagen.

Watts, P. D. 1983. Ecological energetics of denning polar bears and related species. D.Sc. diss., University of Oslo.

Wemmer, C., M. Von Ebers, and K. Scaw. 1976. An analysis of the chuffing vocalization of the polar bear. *Journal of Zoology* (London) 180:425–39.

# Index

INDEX